Freshwater Mussel Propagation for Restoration

Freshwater mussels are declining rapidly worldwide. Propagation has the potential to restore numbers of these remarkable organisms, preventing extinction of rare species and maintaining the many benefits that they bring to aquatic ecosystems. Written by practitioners with firsthand experience of propagation programs, this practical book is a thorough guide to the subject, taking readers through the process from start to finish. The latest propagation and culture techniques are explored as readers follow freshwater mussels through their amazing and complex life cycle. Topics covered include the basics of building a culture facility, collecting and maintaining brood stock, collecting host species, infesting host species with larval mussels, collecting and culturing juvenile mussels, releasing juveniles to the wild, and post-release monitoring. This will be valuable reading for any biologist interested in the conservation of freshwater mussel populations.

MATTHEW A. PATTERSON is a Fish and Wildlife Biologist at the United States Fish and Wildlife Service's National Conservation Training Center in West Virginia. Since arriving at the NCTC, Matthew has created three formal training courses in freshwater mussels, Conservation Biology of Freshwater Mussels, Freshwater Mussel Identification, and Freshwater Mussel Propagation for Restoration.

RACHEL A. MAIR is a Fisheries Biologist with the United States Fish and Wildlife Service. She directs operations for the mussel program at the cooperative Virginia Fisheries Aquatic Wildlife Center at Harrison Lake National Fish Hatchery. Rachel has over nineteen years of experience in freshwater mussel propagation and has worked with over 90 species.

NATHAN L. ECKERT is a Mussel Biologist with the United States Fish and Wildlife Service at the Genoa National Fish Hatchery. Nathan has over 16 years of experience raising mussels from the Mississippi and Tennessee River basins.

CATHERINE M. GATENBY is a Senior Fish Biologist with the United States Fish and Wildlife Service at the Lower Great Lakes Fish and Wildlife Conservation Office. She managed the Freshwater Mussel Propagation Program at the White Sulphur Springs National Fish Hatchery for ten years.

TONY BRADY is a Deputy Project Leader with the United States Fish and Wildlife Service at the Welaka National Fish Hatchery.

JESS W. JONES is a Restoration Biologist with the United States Fish and Wildlife Service, based at Virginia Tech University, Blacksburg. He is also an Associate Professor at Virginia Tech.

BRYAN R. SIMMONS is a Fish and Wildlife Biologist with the United States Fish and Wildlife Service at the Missouri Ecological Services Field Office.

JULIE L. DEVERS is a Fish Biologist with the United States Fish and Wildlife Service at the Maryland Fish and Wildlife Conservation Office.

Freshwater Mussel Propagation for Restoration

MATTHEW A. PATTERSON
US Fish and Wildlife Service,
National Conservation Training Center,
Shepherdstown, WV

RACHEL A. MAIR
US Fish and Wildlife Service,
Harrison Lake National Fish Hatchery,
Charles City, VA

NATHAN L. ECKERT
US Fish and Wildlife Service,
Genoa National Fish Hatchery,
Genoa, WI

CATHERINE M. GATENBY, PH.D.
US Fish and Wildlife Service,
Lower Great Lakes Fish and Wildlife Conservation Office,
Basom, NY

TONY BRADY
US Fish and Wildlife Service,
Welaka National Fish Hatchery,
Welaka, FL

JESS W. JONES, PH.D.
US Fish and Wildlife Service,
Virginia Field Office,
Blacksburg, VA

BRYAN R. SIMMONS
US Fish and Wildlife Service,
Ecological Services & Department of Biology,
Missouri State University,
Springfield, MO

JULIE L. DEVERS
US Fish and Wildlife Service,
Maryland Fish and Wildlife Conservation Office,
Annapolis, MD

CAMBRIDGE
UNIVERSITY PRESS

CAMBRIDGE
UNIVERSITY PRESS

University Printing House, Cambridge CB2 8BS, United Kingdom

One Liberty Plaza, 20th Floor, New York, NY 10006, USA

477 Williamstown Road, Port Melbourne, VIC 3207, Australia

314–321, 3rd Floor, Plot 3, Splendor Forum, Jasola District Centre,
New Delhi – 110025, India

79 Anson Road, #06-04/06, Singapore 079906

Cambridge University Press is part of the University of Cambridge.

It furthers the University's mission by disseminating knowledge in the pursuit of
education, learning, and research at the highest international levels of excellence.

www.cambridge.org
Information on this title: www.cambridge.org/9781108445313
DOI: 10.1017/9781108551120

First published 2018

A catalogue record for this publication is available from the British Library.

Library of Congress Cataloging-in-Publication Data

Names: Patterson, Matthew A., author.
Title: Freshwater mussel propagation for restoration / Matthew A. Patterson, U.S. Fish
  and Wildlife Service, National Conservation Training Center, Shepherdstown, WV,
  Rachel Mair, U.S. Fish and Wildlife Service, Harrison Lake National Fish Hatchery,
  Charles City, VA, Nathan Eckert, U.S. Fish and Wildlife Service, Genoa National Fish
  Hatchery, Genoa, WI, Catherine M. Gatenby, Ph.D., U.S. Fish and Wildlife Service,
  Lower Great Lakes Fish and Wildlife Conservation Office, Basom, NY, Tony Brady, U.S.
  Fish and Wildlife Service, Welaka National Fish Hatchery, Welaka, FL, Jess W. Jones,
  Ph.D., U.S. Fish and Wildlife Service, Virginia Field Office, Blacksburg, VA, Bryan R.
  Simmons, U.S. Fish and Wildlife Service, Ecological Services & Department of Biology,
  Missouri State University, Springfield, MO, Julie L. Devers, U.S. Fish and Wildlife
  Service, Annapolis, MD.
Description: Cambridge, United Kingdom; New York, NY: Cambridge University Press,
  2018. | Includes bibliographical references and index.
Identifiers: LCCN 2017058820 | ISBN 9781108445313 (paperback : alk. paper)
Subjects: LCSH: U.S. Fish and Wildlife Service. | Freshwater mussels—Reproduction.
Classification: LCC QL430.6 .P38 2018 | DDC 594/.4—dc23 LC record available at
  https://lccn.loc.gov/2017058820

ISBN 978-1-108-44531-3 Paperback

Contents

Color plates are found between pages 48 and 49.

Foreword

The science-based propagation of North American freshwater mussels was first undertaken just over a century ago. The goal at that time was to sustain the pearl button industry, which was then valued at today's equivalent of $260 million annually. W.C. Curtis and G. Lefevre, academic biologists at the University of Missouri, carried out seminal studies on mussel reproduction, ecology, and captive culture in an attempt to make the industry sustainable. Their efforts led the US Bureau of Fisheries, forerunner of today's US Fish and Wildlife Service, to establish the Fairport Biological Laboratory in 1914 (Pritchard, 2001). Over the next 20 years, work at the Fairport Laboratory established much of the basic knowledge and many of the methods that underpin our efforts to conserve endangered mussels today. The reports of Lefevre and Curtis and their contemporaries are wonderful in detail and insight, and they are recommended reading for any budding mussel culturist (e.g. Lefevre and Curtis, 1912; Coker *et al.*, 1921).

The modern motivation for mussel culture is quite different from that of the button days. Over the past 50 years there has been a transition in the conservation ethos, from a utilitarian view to one that recognizes the intrinsic value of nature and biodiversity. In terms of fisheries management, a tipping point was reached in the 1960s. For example, in 1962, the USFWS poisoned 19 km of the Green River in Utah and Wyoming with rotenone. The operation was intended to remove native fish species and carp in order to make room for rainbow trout, an act then routine in most respects. Just 4 years later, in 1966, the same biologist that oversaw the rotenone operation was heading a new Committee on Rare and Endangered Wildlife Species, formed by Secretary Stuart Udall, and responsible for protecting the native fish (Wiley, 2008; Halverson, 2010). This philosophical shift was codified by the passage of the Endangered Species Act in 1973. Mussels of course, were among the first species listed under the ESA, and most of the accomplishments reported in this book have been carried out under its auspices.

Over the past 25 years, we have made tremendous strides in the culture of freshwater mussels. Simple rearing systems and artificial diets enable efficient laboratory and hatchery production of juveniles, and systems for holding in raceways, rivers, and ponds allow grow-out with natural food. Continuing advances are improving survival and growth rates. Since the 1990s, more than 20 North American resource agencies, universities, and zoos have developed conservation-oriented mussel culture programs. To date, nearly half of North American species have been metamorphosed on fish hosts or *in vitro*, about a third cultured to several months of age, and many to sexual maturity. Carefully planned restoration efforts, informed by genetic studies and improved understanding of habitat requirements, are underway for many critically endangered species.

Equally as important as restoration is the use of propagated mussels for research, particularly in toxicology. Studies of mussel sensitivity to important pollutants have informed regulatory agencies, and are leading to tightening of federal and state water quality criteria. In 2013, the USEPA revised the allowable levels of ammonia in fresh water downward by more than half, a change which is driving the upgrading of sewage treatment facilities nationwide. This revision was the culmination of more than a decade of work by many researchers and was made possible by the culture of "sacrificial clams" for study. Other criterion revisions are likely to result. Mussels are among the top five most sensitive species for alachlor, ammonia, potassium chloride, sulfate, and nickel, and among the top ten most sensitive species for sodium chloride and copper (Wang *et al.*, 2011, 2016). Laboratory toxicology results can be extended to the field, supporting the prosecution of polluters by demonstrating the mechanism of effects. Propagated juveniles can be caged on-site, acting as sentinels or testing the suitability of conditions for population restoration.

Many challenges remain, and I look forward to revisions of this book in the future as these challenges are met. Particular needs include continued research on physical and dietary requirements of juvenile mussels, particularly habitat specialists. The ecology and limiting factors of early juveniles are still poorly known. The potential positive and negative genetic impacts of stocking programs should be investigated experimentally. The basic needs recognized a century ago have not yet been met: we need to prevent the loss of abundance as well as diversity. Ecosystem services depend mainly on abundance, whereas the preservation of rare species depends on population restoration. Stream ecosystems are increasingly fragmented. As populations become smaller and more isolated, their vulnerability will increase. Impacts of pollution, invasive species, and climate change are unlikely to abate. The capacity of resource agencies to culture, stock, and monitor populations is a necessity to manage mussels and prevent further extinctions over coming decades.

Chris Barnhart, Ph.D.
Distinguished Professor of Biology
Missouri State University,
Springfield, MO

Acknowledgements

The other authors thank Matthew Patterson for compiling and editing the whole volume.

MATTHEW PATTERSON

I first would like to thank all of my co-authors for their hard work on the course outline, course description, course manual, and final book manuscript. Without your dedication to freshwater mussel conservation, this book would not have been possible.

I would like to thank colleagues that provided editorial comments on this book including Wendell Haag, Brian Watson, Paul Johnson, Kelly Bibb, Chris Eads, Mark Hove, Bernard Sietman, and Jim Siegel.

I also would like to thank all of the folks that provided photographs, tables, and figures for the book, including Chris Barnhart, Ryan Hagerty, Angela Boyer, Bruce Young, Kristin Simanek, Bernard Sietman, Rachel Mair, Molly Webb, Jonathan Wardell, Bryan Simmons, Nathan Eckert, Rachael Hoch, Amanda Wood, Megan Bradley, Amy Maynard, Jaclyn Zelko, Frankie Thielen, Beth Glidewell, Paul Johnson, Julie Campbell, Amy Thompson, Wendell Haag, Janet Butler, Gary Wege, Sara Weglein, Jeremy Tiemann, Janet Clayton, and Michael Odom.

Thanks also to Karene Motivans and Jim Siegel for their valuable assistance in the final phase of this project and to all my coworkers at the National Conservation Training Center (USFWS) for supporting my effort to publish this course manual.

A big thank you goes out to Dr. Chris Barnhart for writing the Foreword and for providing numerous photographs for the book. It is hard to imagine where freshwater mussel propagation and freshwater mussel conservation in general would be today without your tireless efforts.

I also would like to send a very special thank you to the five gentlemen who hooked me on aquatic biology and freshwater mussels, Dr. Rick Kopp (Georgetown College), Dr. Guenter Schuster (Eastern Kentucky University), Ron Cicerello (Kentucky State Nature Preserves Commission), Ellis Laudermilk (Kentucky State Nature Preserves Commission), and Dr. Richard Neves (Virginia Polytechnic Institute and State University). Your inspiration, guidance, and support have been critical throughout my career.

Finally, I would like to thank my family. Uncle John Fleischauer for the long walks at the Patterson farm discussing the finer points of writing, my father (Larry "Doc" Patterson), mother (Mary Patterson) and brother (Michael Patterson) for exposing me to the great outdoors and my lovely wife Diana and son Mazi for all of your love and support. You guys are the greatest!

CATHERINE GATENBY

I would like to thank Dr. Danielle Kreeger for her guidance in studying bivalve feeding physiology and developing freshwater mussel culture technology, Dr. David Orcutt for expertise in designing production-scale algae cultures and analyzing algae biochemistry, Dr. Bruce Parker for guidance on selecting algae high in lipids and culture techniques, and Dr. Richard Neves for vision and funding research on propagation of mussels.

JULIE DEVERS

I would like to thank Dr. Richard Neves for increasing knowledge and awareness of the conservation needs of freshwater mussels.

NATHAN ECKERT

Let me start by thanking Matthew Patterson for all of his hard work bringing this book together, it would not have arrived here without him. I am also grateful for the rest of my co-authors and their contributions to the project. I would like to thank my parents for always sending me outside and allowing me to play in the water. I owe a debt of gratitude to my science teachers Max McGee, Jim Jacobs, Paula Rose, Dr. Charlie Hunter, Gene Young, and Dr. Pat Ross for fostering my interest in science. I want to thank Ryan Waters, Joe Ferraro, and Mike Pinder for shaping who I would become as a professional biologist. I must also thank Dr. Chris Barnhart for introducing me to the wonderful world of freshwater mussels and for all that he invested in me as a student and young biologist. Finally, and most importantly, I wish to thank my wife Shelley for her love and support and for accompanying me on this great journey all over the country looking for these interesting little rocks with guts.

BRYAN SIMMONS

I would like to thank all the dedicated mussel biologists who have helped contribute to our understanding of freshwater mussel ecology and conservation. I would also like to thank the co-authors, Chris Barnhart and his students, Andy Roberts, Josh Hundley, Scott Faiman, Steve McMurray, and my former co-workers for their collaboration and friendship. Above all, I would like to thank my family for their endless patience and support.

RACHEL MAIR

I would first like to thank all of the authors who came together to write this book, especially Matthew Patterson. Next, thank you to the people that started me on my mussel journey whose passion for these animals was truly contagious and got me hooked: Bill Henley (Virginia Polytechnic Institute and State University), Dr. Richard Neves (Virginia Polytechnic Institute and State University-retired), Jess Jones (United States Fish and Wildlife Service), Steve Ahlstedt (United States Geological Survey-retired), and many others on the river. I would also like to thank my current and former supervisors Michael Odom and Catherine Gatenby. Additionally, I would also like to thank all of my co-workers through the years; especially Brian Watson (VDGIF) for his hard work establishing the mussel program at Harrison Lake.

Lastly I would like to thank my parents, John and Madeline, my wonderful husband (David Garst) and my sweet son Andrew for all your love, support, and patience.

TONY BRADY

I would like to acknowledge and thank first of all, my family who has put up with my love of freshwater mussels for the past 20 years. Thank you to my major professor Dr. Jim Layzer who gave me the opportunity to study mussels under his guidance and for his encouragement that helped me to become the malacologist I am today. Finally, I would like to thank my co-authors for allowing me to be a part of this book, their respect and friendship I will cherish always.

JESS JONES

First, I would like to acknowledge my graduate advisor Dr. Richard Neves, US Geological Survey (Retired), for giving me the opportunity to work with and study mussels while an employee and student at Virginia Tech from 1998–2009 and the role he has played in establishing our field of mussel propagation and culture. Dr. Neves' tireless efforts from 1978–2008 initiated the present era of mussel research and conservation in the United States, broadly influencing the scientific methodologies in use today, shaping key hypotheses concerning mussel life history and the reasons for their decline, and most importantly, developing and advocating for the use of propagation and culture technology as a primary conservation strategy for mussels in this country and abroad. In 2002, he established the Freshwater Mollusk Conservation Center at Virginia Tech, the first facility of its kind in the United States, which served as the model for many other facilities that followed. His vision, dedication, generosity, and kindness influenced a generation of biologists, many of whom helped write the chapters of this book. I also would like to thank Matthew Patterson for inviting me to participate as a book co-author and to serve as an instructor in 2015 for the mussel propagation course taught at the USFWS National Conservation Training Center, Shepherdstown, West Virginia.

Note from the Authors

The findings and conclusions in this book are those of the authors and do not necessarily represent the views of the US Fish and Wildlife Service.

Global declines in freshwater mussel populations make freshwater mussel conservation and propagation a global effort. With that said, high levels of species diversity combined with high levels of impairment has led to a lot of research on freshwater mussel propagation methods in the United States. As a result, this book relies heavily on both published and unpublished methods developed in the United States. To the best of their ability, the authors also have reported on methods currently in use in other parts of the world (e.g. efforts in Europe to propagate the freshwater pearl mussel, *Margaritifera margaritifera*). At the time this book was written, the authors were unable to find information on culture techniques being used in South America, although we believe efforts are currently underway in Brazil.

Note from the Authors

1 Why Propagate Freshwater Mussels?

Matthew A. Patterson, Jess W. Jones
and Catherine M. Gatenby

1.1 GLOBAL CONSERVATION STATUS

Freshwater mussels have successfully colonized every continent on Earth except Antarctica (Bogan, 2008); yet, nearly half of the global freshwater mussel species are listed as imperiled (IUCN, 2016). Without concerted efforts toward conservation, Ricciardi and Rasmussen (1999) project that 127 freshwater mussel species could go extinct by 2099. In North America, 74% of the known 300 species are imperiled, 220 of which are listed as endangered, threatened or of special concern in the United States, and at least 35 species are considered extinct (Williams *et al.*, 1993; Neves *et al.*, 1997) (Figure 1.1). Of the 16 species recognized in Europe, 3 are critically endangered, 2 are vulnerable, and 5 species are near threatened (Lopes-Lima *et al.*, 2017). Indeed, all European *Margaritifera* species are listed as "fauna requiring special measures to be taken for their protection" under the provisions of the Bern Convention on the Conservation of European Wildlife and Natural Habitats. Bauer (1986) reported that southern Europe's freshwater pearl mussel (*Margaritifera margaritifera*) had declined markedly with only 2 of the 12 rivers supporting stable populations. Freshwater pearl mussel populations in Russia follow a similar trend, with 70% of historic populations in southern Russia likely extirpated (Popov and Ostrovsky, 2014). Geist (2010) cautions that an ongoing lack of recruitment may lead to further declines in Europe's freshwater mussel populations. The conservation status of freshwater mussels in South America is less well known, but many species are declining (Pereira *et al.*, 2014). Ninety-three percent of the freshwater bivalves in Uruguay are priorities for conservation

FIGURE I.I The clubshell (*Pleurobema clava*) collected from the Allegheny River (Forest County, Pennsylvania). The clubshell is listed as federally endangered in the United States under the Endangered Species Act of 1973.
Photo: Ryan Hagerty, USFWS.

(Clavijo *et al.*, 2010). One percent of Brazil's freshwater bivalves are listed as critically endangered, 10% endangered, 9% vulnerable, and 37% are in need of further evaluation for potential listing (Pereira *et al.*, 2012). One species is listed as vulnerable in Columbia (Ardila *et al.*, 2002) and eight species are listed as endangered by the Ministerio de Agricultura Ganaderia in Paraguay. There also are data gaps in our knowledge of freshwater mussels in Africa and Australasia (Seddon *et al.*, 2011, Walker *et al.*, 2014). Seddon *et al.* (2011) estimates that 25% of the species in Africa are extinct, critically endangered, endangered, vulnerable, or near threatened, and an additional 25% are lacking sufficient data to assess their status. Of the 32 species found in Australasia, 7 are listed by the International Union for the Conservation of Nature or under national or state legislation, and Walker *et al.* (2014) believe that a much-needed systematic revision

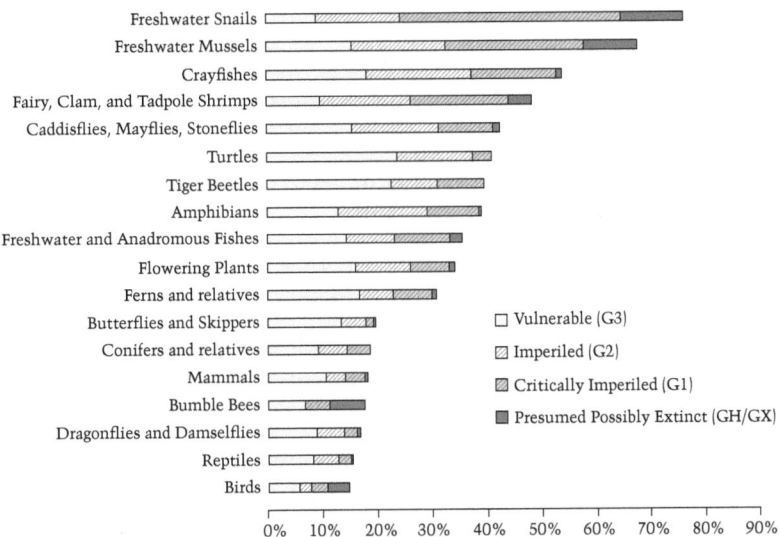

FIGURE I.2 Percentage of species listed as vulnerable, imperiled, critically imperiled, and presumed possibly extinct by faunal and floral group in North America. The figure clearly shows that freshwater species are more imperiled than their terrestrial counterparts. Graphic courtesy of NatureServe and adapted by Kristin Simanek, USFWS.

will likely increase this number. Little or no information is available on the conservation status of freshwater mussels in China. In 2008, however, the Ministry of Environmental Protection categorized 45% of the major river reaches as either moderately or badly polluted, with all freshwater fishes eliminated from a full 5% of the length of Chinese rivers as a direct result of pollution (Dudgeon, 1999). It is likely, therefore, that impacts on freshwater mussels in China are similar to other parts of the world.

Mussels are certainly not the only imperiled group of animals in freshwater ecosystems (Figure 1.2). Strayer (2006) estimates that approximately 12 000 species of freshwater invertebrates are either extinct or imperiled globally. Of the 703 freshwater snail species (Gastropoda) in the United States and Canada, 74% are listed as vulnerable, threatened or endangered and 67 species are considered extinct or possibly extinct (Johnson *et al.*, 2013). Richman *et al.* (2015) estimate that 32% of the nearly 600 species of crayfish worldwide are in danger of extinction. In the United States and Canada, 174 of the 363 crayfish species (48%) are listed as endangered, possibly extinct, threatened or vulnerable (Taylor *et al.*, 2007). Freshwater vertebrates are not immune to these declines. Between 1898 and 2006, 57 species of freshwater fishes in North America went extinct (Burkhead, 2012). Of the remaining 700 species of fishes in North America, 39% are listed as imperiled (Jelks *et al.*, 2008). The global conservation status of freshwater fishes is more difficult to assess due to significant data gaps. Only about 5800 of the 15 570 described freshwater fish species (37%) had been assessed as of 2011, with 30% of the 5800 listed as extinct, extinct in the wild or threatened with extinction (Carrizo *et al.*, 2013). These statistics paint a clear picture that freshwater systems in North America and around the world are in trouble. In fact, extinction rates in freshwater ecosystems appear to be five times higher than extinction rates in terrestrial systems and even rival extinction rates for the tropical rainforests (Ricciardi and Rasmussen, 1999).

1.2 CAUSES OF THE DECLINE

As with most imperiled species, it can be difficult to point to a single cause for the decline of freshwater mussel populations. Authors from around the world have cited similar impacts to freshwater mussel populations, from habitat loss and alteration, commercial harvest, pollution, loss of fish hosts, to competition with invasive species (Bauer, 1988; Lydeard *et al.*, 2004; Strayer *et al.*, 2004; Nobles and Zhang, 2011; Haag, 2012; Lopes-Lima *et al.*, 2014; Pereira *et al.*, 2014). Downing *et al.* (2010), however, found few research papers that provided evidence of a direct causal link between mussel declines and any specific impact. The most frequently cited cause for the decline of freshwater mussel populations, and numerous other faunal groups around the world, is habitat loss. Human activity has left very few rivers on Earth unimpacted (Vorosmarty *et al.*, 2010). The construction of nearly 1 million dams globally (Jackson *et al.*, 2001) has resulted in the loss of many natural, free-flowing river segments, negatively impacting flow rates, sediment loads, temperature regimes, and dissolved oxygen upstream and downstream of the dams. In North America, only 40 river segments larger than 200 km (125 miles) in length are still free-flowing (Benke, 1990). By the mid-1940s, the upper Mississippi River and the entire length of the Tennessee River were controlled by dams (Etnier and Starnes, 1993; Anfinson, 2003). One stark example of the impacts of impoundments on freshwater mussels comes from Fort Loudoun Reservoir on the Tennessee River. Prior to impoundment, this stretch of the Tennessee River supported 64 species of freshwater mussels (Ortmann, 1918). After impoundment, only 4 species remained (Isom, 1971). Impoundments can be tied directly to the extinction of at least 12 species of freshwater mussels (Haag, 2012).

Beginning in the 1960s, however, mussel populations began to decline in unimpounded and seemingly healthy streams (Haag, 2012). The Embarrass River, for example, experienced an 86% decline in the freshwater mussel fauna, despite the river being classified as one of Illinois' outstanding streams (Cummings *et al.*, 1988). Crashes to other local mussel faunas were documented all over the United States, including the states of Georgia, Iowa, Kansas, Kentucky, Michigan, Tennessee, Texas, and Virginia (Distler and Bleam, 1995; Howells *et al.*, 1997; Evans, 2001; Haag and Warren, 2004; Poole and Downing, 2004; Hanlon *et al.*, 2009; Morowski *et al.*, 2009; Jones *et al.*, 2014). While many of these rivers support healthy assemblages of fishes and aquatic insects, extant mussel populations are comprised almost entirely of old, relict individuals (Haag, 2012; Strayer and Malcom, 2012). Lack of recruitment had led to the disappearance of short-lived species followed by a gradual decline of the longer-lived species (Haag, 2012). Strayer and Malcom (2012) were interested in identifying possible causes for mussel recruitment failure in rivers in southeastern New York. They found no relationship between recruitment failures and fine sediment, interstitial oxygen concentration, fish host abundance, or crayfish predator abundance. They did, how-ever, show that concentrations of un-ionized ammonia greater than 0.2 mg N/L in the interstitial water were correlated with recruitment failure. Recent laboratory research also has shown that the juvenile life stage of freshwater mussels is highly susceptible to environmen-tal contaminants. In fact, juvenile mussels may be an order of magni-tude more susceptible than the standard test organisms used by the United States Environmental Protection Agency for contaminants like ammonia and copper (Wang *et al.*, 2007a, 2007b). Additional research is needed to better understand the causes of these fresh-water mussel declines. If the causes are not identified and ameliorated, recovery efforts are likely to fail.

1.3 ECOLOGICAL SIGNIFICANCE OF FRESHWATER MUSSELS

Healthy freshwater mussel beds can make up 50–90% of the benthic biomass in streams, in some cases exceeding the biomass of all other benthic species combined by an order of magnitude (Negus, 1966; Layzer *et al.*, 1993; Strayer *et al.*, 1999). Because an organism's (or group of organisms') contribution to ecological processes is directly proportional to their biomass (Strayer *et al.*, 1999; Vaughn *et al.*, 2004), the severe decline in mussel populations described above is likely having a significant impact on freshwater ecosystems. Indeed, a growing understanding of the ecology and physiology of freshwater mussels indicates they play a significant role in structuring food webs and providing ecological functions important to maintaining the over-all health of the ecosystem (Zimmerman and Szalay, 2007; Vaughn, 2010; Allen and Vaughn, 2011; Allen *et al.*, 2012; Atkinson *et al.*, 2013, 2014a; Strayer, 2014; Atkinson and Vaughn, 2015). For exam-ple, freshwater mussels, like many marine bivalves, are extremely efficient filter feeders. Large mussel beds are capable of filtering the entire volume of water passing over the bed at any given time (Welker and Walz, 1998; Vaughan *et al.*, 2004). Exponential declines in phyto-plankton biomass in the River Spree in Germany were attributed to filtration by dense freshwater mussel beds (350 mussels/m²; Welker and Walz, 1998). Additionally, as mussels convert filtered organic material into excretory products, nutrients are transferred from the water column to the benthos (Spooner and Vaughn, 2006). Where nutrients are limiting, mussel excreta can support the rest of the food web, leading to increases in benthic algae, macroinvertebrates, and fish (Allen *et al.*, 2013; Atkinson *et al.*, 2014b). In the Kiamichi River in southeastern Oklahoma, benthic areas around live mussel beds had higher invertebrate abundance and organic matter concentrations than areas of the river with no mussels (Spooner and Vaughn, 2006). Mussel-provided nutrients also can alter algal composition, decreasing blue-green algae and increasing water quality (Atkinson *et al.*, 2013). The burrowing behavior of freshwater mussels can increase water and oxygen penetration through the sediment, as well as release nutrients from sediments and stabilize river substrates (Matisoff *et al.*, 1985; McCall *et al.*, 1995; Zimmerman and de Szalay, 2007; Allen and Vaughn, 2011). Live mussels also are an important food resource for fishes, mammals, and birds, and dead shells are a source of calcium as well as habitat for some aquatic organisms (Vaughn, 2010). Finally, intact mussel assemblages can improve conditions for rare species (Spooner and Vaughn, 2009).

1.4 RECOVERY EFFORTS

The high level of imperilment in global freshwater mussel populations combined with their important function as "ecosystem engineers" is causing great concern among scientists, prompting the creation of freshwater mussel conservation programs around the world (Lydeard *et al.*, 2004; Strayer *et al.*, 2004). This book is designed to introduce the reader to one aspect of these multi-faceted conservation programs, the propagation and stocking of freshwater mussels.

As early as 1899, biologists with the United States Bureau of Fisheries' Fairport Biological Station began developing mussel propagation techniques to supplement populations declining as a result of harvest for the pearl button industry (Pritchard, 2001). After extensive work on host fish relationships and juvenile mussel culture, the Fairport Station closed in 1933. In the 1960s, widespread reports of dramatic declines in freshwater mussel populations renewed research into mussel propagation technology. This new research has resulted in vast improvements in life history information and rearing practices, as well as the development of mussel propagation programs in the United States, Europe, and other countries (Hastie and Young, 2003; Neves, 2004; Thomas *et al.*, 2010; Haag and Williams, 2014). In the United States, the primary goal has been species recovery. Management objectives in many Endangered Species Recovery Plans and State Wildlife Action Plans include restoring viable populations to a significant portion of the species historic range, restoring resilience to environmental impacts, and preventing new species from being listed under the Endangered Species Act of 1973. Several propagation programs also are working to restore mussel beds impacted by chemical spills or other instream activities, like bridge replacements (Morrison *et al.*, 2013; Lane *et al.*, 2014). There is a growing demand, however, for propagation programs aimed at restoring complete mussel assemblages, including both common and rare species. Restoring the ecosystem goods and services provided by freshwater mussel beds is likely fundamental to the long-term viability of mussel populations (Vaughn and Spooner, 2006). In fact, endangered species recovery efforts will likely be more successful if the complete mussel assemblage is restored (Haag, 2012). Mussel propagation facilities also are frequently tasked with producing mussels for research and serving as centers for holding captive populations of species on the brink of extinction (i.e. ark populations). Finally, propagation facilities provide an ideal environment to inform the public about the importance of freshwater mussels.

Restoring mussel habitat and improving water quality also must be key elements to any multi-faceted conservation effort. Unfortunately, once populations have declined, it can take decades for freshwater mussels to recolonize restored habitat, due to their complex life history. Consequently, propagation programs are oftentimes necessary to prevent extinction prior to habitat restoration or to jump-start species recovery after the habitat is restored. As a result, propagation is likely to remain a key conservation management strategy for restoration and recovery of freshwater mussels into the future. Propagation of freshwater mussels, however, should not be taken lightly and any facility that is considering starting a propagation program should carefully consider the following questions.

I.5   QUESTIONS TO CONSIDER BEFORE STARTING A FRESHWATER MUSSEL PROPAGATION PROGRAM

I.5.I   *Why Are We Propagating Freshwater Mussels and is Propagation the Best Restoration Strategy?*

The first question to answer before starting a mussel propagation program is "Why are we propagating freshwater mussels and is propagation the best restoration strategy?" Defining clear goals and objectives for the program is critical. Juvenile mussels can be produced for basic life history research, toxicity testing, restoration of historical populations and the ecosystem services they provide, and recovery of imperiled species. A planning team that includes local experts and other stakeholders can help identify the primary goals and objectives. Working with stakeholders early in the process also can help ensure agreement on a path forward and commitment to implementing the planning team's recommended actions.

If restoration and recovery are identified as one of the program goals, the planning team must first determine if propagation is the best restoration strategy for any given mussel species or population. Controlled propagation is not a substitute for addressing factors responsible for an endangered or threatened species' decline (USFWS and NMFS, 2000), and should be considered as a last resort. If the factors responsible for the decline are not addressed, propagation efforts are likely to fail. The United States Policy Regarding Controlled Propagation of Species Listed under the Endangered Species Act states that the first priority is to recover wild populations in their natural habitat. Before collecting a single gravid mussel or building a single juvenile culture system, consider any and all feasible alternatives to propagation, such as habitat restoration and water quality improvements. Suitable habitat that provides the necessary resources for growth and reproductive success is critical. Stocking mussels in degraded habitats will not help recover the species and ultimately will be a waste of money. Investing in long-term solutions like habitat restoration, while simultaneously working to improve propagation technology might be a better strategy. Once the habitat is restored, a stocking program can help meet restoration objectives for the species.

Unfortunately, many freshwater mussel populations have declined to the point that the time for last resorts has arrived. In some cases, populations are so low that extinction is imminent. The purple catspaw (*Epioblasma obliquata obliquata*) provides an excellent example (Figure 1.3). Once widespread in the southern Ohio River basin, the purple catspaw is now considered one of the rarest mussel species in North America. Initially thought to be functionally extinct, a breeding population was discovered in Killbuck Creek, Ohio, in 1994. Since the discovery of this new population, Killbuck Creek has become degraded to the point that drastic measures are now necessary to prevent extinction of the purple catspaw. Efforts to address the factors responsible for the decline of catspaw are currently underway, including fencing out livestock to restore stream banks, prevent sedimentation and restore the stream channel. Unfortunately, this species could go extinct before the benefits of habitat restoration are fully realized. Thus, propagation will play a critical role in preventing extinction of this species.

Taking no conservation action also needs to be carefully considered. If any proposed conservation measure, including propagation, has the potential to do more harm to imperiled species than good, it may be better to do nothing. If all feasible alternatives have been exhausted and it becomes clear that propagation is needed to recover a species or prevent extinction, the remaining questions should be addressed before propagation begins. All of these questions are

FIGURE 1.3 Fresh-dead shells of a male (top) and female (bottom) purple catspaw (*Epioblasma obliquata obliquata*).
Photo: Angela Boyer, USFWS. (A black-and-white version of this figure will appear in some formats. For the color version, please refer to the plate section.)

important, so the order of the questions is not intended to indicate relative importance.

1.5.2 Do We Have a Propagation and Species Restoration Plan?

A detailed Propagation and Species Restoration Plan (Plan) that defines clear goals and objectives for the program is critical. At a minimum, the Plan should include details about priority species, culture techniques, release sites, endpoints for success, etc. The questions identified in this chapter will help with preparation of the Plan, but forming a planning team of experts, partners, and other stakeholders to help write the draft Plan is highly recommended. An advisory committee to review and provide comments on the draft Plan also is recommended. A thorough review process can help ensure agreement on the path toward recovery, as well as commitment to implementing the recommended actions contained in the Plan.

In addition to working closely with a review team, propagation facilities should be coordinating all restoration activities (propagation, stocking, habitat restoration, translocation, etc.) across the species entire range. Coordination with other agencies working on the same species can help: (1) prevent catastrophic losses associated with "putting all the eggs in one basket", (2) ensure stocking programs meet agreed-upon population objectives for the entire species, and (3) minimize harm to existing mussel communities and brood stock populations from over-collecting. Coordination across a species entire range will require detailed record-keeping, data management, and publication of all results.

1.5.3 Have Suitable Stocking or Release Sites Been Identified?

It is crucial that suitable release sites are identified before propagation is considered as a viable restoration option. Scrambling to find suitable release sites after juvenile mussels have been produced and cultured to a suitable size for release will likely lead to an inappropriate or ill-advised release. A clear set of criteria should be established for assessing the suitability of the site for supporting all life stages of freshwater mussels, including substrate conditions, the presence of suitable fish hosts, food availability, and water quantity and quality. All potential release sites should be vetted by local experts and the advisory committee. If suitable habitat is not present, stocking efforts are likely to fail (George et al., 2009).

Releasing cultured juveniles in cages at the proposed release site is one useful technique for assessing suitability (see Section 6.4.5). If growth and survival of juvenile mussels in the cages are poor relative to reference sites, the release site may be unable to support mussels in its current state. Habitat restoration and water quality tests should be completed before mussels are propagated for release or a new release site should be selected.

I.5.4 Have Clear Population Objectives Been Established and Have Optimal Stocking Numbers, Frequency, and Duration Been Established to Meet Those Objectives?

Research on what is a viable population size for mussels is lacking. Even in those cases where demographic data is available, it may not be applicable to other species. Propagation plans, however, should aim to include population objectives using the best information available and expert knowledge of the species. The number of progeny to stock, including possible sex ratios, how often stockings will occur (annual, biannual, or other), the time of year (spring, summer, or fall) stockings will occur, and determining when recovery has been achieved and stockings are no longer needed, would be based on the population objectives. Once a sustainable population with natural recruitment is established, stocking efforts can either cease or move to another release site.

If a mussel species has intrinsically high population growth rates, frequent stocking over a long period of time may be unnecessary for recovery (Canessa et al., 2014). The fragile papershell (*Leptodea fragilis*), a very thin-shelled, fast-growing species that has been shown to become sexually mature before age one (Haag and Rypel, 2011), may not require frequent stocking. For species with intrinsically low population growth rates, regular stocking over a long period of time may be necessary to avoid extinction. The dromedary pearlymussel (*Dromus dromas*) is a slow-growing and -maturing species that matures around age 8 (Jones et al., 2004), and therefore may require more frequent stocking over a longer period of time.

Additionally, a balance between stocking enough mussels to sustain a population in the wild and stocking too many mussels must be carefully examined. Potential consequences of over-stocking a population with cultured mussels can include reduced genetic variability, swamping the gene pool with an influx of genetically similar individuals, as well as exceeding the carrying capacity at the release site (George et al., 2009). Finding this balance will depend on the life history strategy and population dynamics of the mussel species being stocked. For freshwater fishes, George *et al.* (2009) recommend stocking small numbers of individuals at multiple locations for species that are poor dispersers (e.g. darters, madtoms, etc.). For strong dispersers, they recommend stocking large numbers of individuals at a single location. In the case of mussels, it would be wise to take dispersal rates into account as well, especially dispersal rates of the host species.

Determining optimal stocking numbers and frequency is complex and would benefit greatly from a combination of population modeling and formal decision-making processes (Jones *et al.*, 2012; Converse *et al.*, 2013; Canessa *et al.*, 2014). Be sure to engage the advisory team as well as all partners and stakeholders during the decision-making process to increase buy-in and to ensure that all potential actions have been considered. Once the optimal stocking numbers have been established, there is a chance a propagation facility could end up with surplus juvenile mussels. As dedicated conservationists and avid mussel biologists, the temptation is to release those juveniles, especially for critically endangered species (George *et al.*, 2009). It is prudent, however, that stocking be limited to the number approved in the Plan. Surplus juveniles can still be put to good use for toxicity studies, research, or outreach to the public.

Finally, it is essential to monitor the success of re-established populations. If monitoring indicates the population is still declining, population objectives and optimal stocking numbers in the Plan can always be adjusted.

1.5.5   Have Desired Endpoints to Measure Success Been Identified?

If no endpoints or measures of success have been identified, it will be difficult to evaluate when the species has recovered at a given release site and additional stocking is no longer needed. Agreement on when population objectives have been achieved and when stocking is no longer needed should be included in the Plan. Regardless of the endpoints selected, these measures should be identified early in the propagation planning process. Once recovery of a species or populations can be confirmed, limited resources can be re-directed toward propagation of other priority species.

1.5.6 Could the Health of the Wild Population Used to Collect Brood Stock for Propagation be Adversely Impacted?

The size and health of the wild population used for brood stock will have a dramatic impact on the number of mussels propagated and ongoing stocking efforts (Canessa *et al.*, 2014). Source populations with large numbers of individuals and high rates of recruitment might be able to withstand the removal of large numbers of individuals with minimal impact. Source populations of critically endangered mussels will likely be unable to sustain large removals. In some cases, populations are so low that it can be hard to even find gravid females in the wild. Population objectives and stocking goals need to consider these factors to avoid undue harm to wild brood stock populations.

1.5.7 Will the Project Represent a Re-introduction or an Augmentation?

If the proposed project is a re-introduction, it is important to verify that the species occurred historically at the proposed release site. No matter how endangered a mussel species may be, it is important to never introduce mussel populations outside their historic range. Introductions of this kind can have unintended consequences for the native species at the release site (George *et al.*, 2009). Even if the release site is part of the historic range of the species, it is important to confirm the species has been extirpated from the site. If extirpation is confirmed, other populations in close proximity could still serve as a source for natural recolonization. Remember, natural recovery is preferred.

If the proposed project is an augmentation of an existing population, are we certain that population levels of the species are too low? A central tenant of community ecology is that some species are and always have been rare. A naturally rare species may already be at carrying capacity despite low population numbers. Attempts to increase abundance beyond carrying capacity are likely to fail (Cox, 1994; Aprahamian *et al.*, 2003). If a population is declining rapidly and there is no evidence of recruitment, it is possible that whatever issues caused the decline in the first place are still present and will lead to failure of the augmentation effort. In this case, re-introduction to a new site within the historic range of the species may be a better recovery strategy.

Some researchers have expressed serious concerns with augmentation using captive propagation (Haag 2012; Haag and Williams, 2014). The primary concern is the potential for negative impacts to genetic diversity; however, impacts to the community and the ecology at the release site also are of concern. Fish stocking has been practiced for decades, with impacts on the ecology of the stocked streams only recently understood (Molony *et al.*, 2003). These impacts have been reported at the species, population, and community levels, representing both genetic and behavioral responses (Aprahamian *et al.*, 2003; Molony *et al.*, 2003; Nickum *et al.*, 2004). The risks may depend on the spatial as well as temporal overlap of stocked and wild fish (McMichael and Pearson, 2001). For example, some hatchery-reared steelhead trout (*Oncorhynchus mykiss*) migrated over 12 km upstream into areas containing westslope cutthroat trout (*Oncorhynchus clarki lewisi*) and a threatened stock of bull trout (*Salvelinus confluentus*). McMichael *et al.* (1999) also reported hatchery-reared steelhead displacing pre-existing wild populations of steelhead. Perrier *et al.* (2013) evaluated 25 populations of Atlantic salmon (*Salmo salar*) throughout France to investigate the influence of stocking on the genetic structure in wild Atlantic salmon. They analyzed fish sampled from 1965 to 2006, and found the overall genetic structure among populations had decreased over the period studied. Similar impacts could result from the stocking of freshwater mussels. For example, releasing large numbers of juvenile mussels produced in the laboratory from only a few gravid females could swamp the gene pool with an influx of genetically similar individuals to the population. A single female in the wild, by contrast, may only produce one offspring per year (Haag, 2012). If gravid females are collected from separate drainages with distinct genetic stocks, augmentation also can lead to outbreeding depression, a decrease in fitness of progeny produced from the mating of genetically incompatible parents.

Finally, augmentation projects by definition do not create new populations, a requirement for delisting or downlisting a federally listed species in the United States. While augmentation may be necessary to prevent extinction of critically imperiled species, careful planning is needed to ensure any proposed augmentation project is doing no harm to the existing mussel fauna.

1.5.8 Have Genetic Concerns Been Addressed?

Captive propagation has the potential to have a significant impact on the genetic diversity of mussel populations in the wild. Potential negative impacts could include extinction, loss of within-population genetic variation, loss of between-population genetic variation, and domestication selection (Busack and Currens, 1995). Some mussel species may exhibit significant genetic differentiation across their historic range due to habitat fragmentation or natural isolation. A thorough assessment of this cryptic diversity is critical for the selection of brood stock for propagation. The mixing of stocks that are genetically differentiated and possibly adapted to local conditions could result in reduced fitness of the progeny (Grobler *et al.*, 2006). Jones *et al.* (2006a) addressed genetic concerns in freshwater mussel propagation at great length, but below is an abbreviated list of their recommendations:

(1) Gravid females used for augmentation of an existing population should be collected from the existing population, if possible.
(2) Gravid females used for re-introduction to a species' historic range should be collected from the closest adjacent watershed that is most similar in terms of genetic and ecological characteristics.
(3) If possible, individual gravid females should only be used once for propagation to prevent swamping the release site with genes from a single female.

(4) A maximum number of gravid females to be collected annually from a given source population should be established to avoid negative impacts to natural reproduction. This guideline is especially important for very small populations.

(5) For large populations, greater than 50 gravid females should be used for propagation to adequately capture the genetic diversity in the wild population. For small populations and critically imperiled species, meeting the recommended minimum of 50 gravid females will be difficult if not impossible. In these cases, collect as many gravid females as possible to maximize genetic diversity, while being careful to minimize any negative impacts to the source population.

(6) Evolutionary significant units, subspecies or closely related species should not be mixed to avoid outbreeding depression.

(7) Culture conditions in the laboratory should mimic natural conditions as much as possible to avoid domestication selection.

Each of these recommendations should be carefully reviewed in the Propagation Plan well in advance of collecting gravid females from the wild.

1.5.9 Have Ecological Concerns Been Addressed?

In addition to genetic concerns, there are ecological concerns associated with propagation that need to be considered. The ultimate goal of mussel restoration should be to restore the complete mussel assemblage and all of the ecological services provided by that assemblage, not just recovery of imperiled species. Growth and survival of imperiled species could be negatively affected without the full range of ecosystem services provided by the more dominant species in the assemblage. Spooner and Vaughan (2006) showed that the presence of abundant keystone species in a mussel bed can have positive effects on the growth and survival of less abundant species in the bed. Consequently, restoration of the full range of ecosystem services provided by the mussel bed may be critical to saving imperiled species (Haag, 2012).

Artificially increasing the abundance of an imperiled species above the natural carrying capacity could have negative consequences for other species in the assemblage (see Section 1.5.7). Some mussel species may be naturally rare in a population, but still stable and recruiting. In fact, if the relative abundance of that species is naturally low, it may already be at carrying capacity for that stream and increasing the population size could impact other species. To make matters worse, streams that already support stable and recruiting populations of mussels tend to be our first choice for release sites. Because of impacts to surrounding habitats, these could be the only quality release sites available. Unfortunately, it is these very areas that may be at greatest risk for ecological harm from propagation and stocking (Haag, 2012).

I.5.10 Do We Have the Appropriate Staff Expertise and Facility for Mussel Propagation?

Freshwater mussel identification can be very challenging and it generally takes several years of fieldwork to reliably identify some species. Mussel propagation also takes very specialized knowledge and training in mussel biology, physiology and ecology, host species care and maintenance, and in some cases, algae culture. It may be necessary to hire new staff with mussel propagation experience. If hiring new staff is not feasible and training existing personnel to culture mussels is the only option, hands-on training at an existing mussel propagation facility is highly recommended. There is no replacement for hands-on experience with an experienced mussel propagation biologist.

It is important to assess if a facility has the available space and appropriate water quantity and quality for mussel propagation. Before building a new mussel propagation facility, investigate whether existing propagation facilities can meet the goals identified in the Propagation Plan. If so, the money and staffing resources that would have been used to build a new facility could instead be put toward additional years of juvenile production and release, habitat restoration, monitoring, or research.

1.5.11 *Are Local, State or Federal Government Permits*
Required to Propagate Freshwater Mussels
and Have You Obtained the Necessary Permits?

In the United States, Section 10(a)(1)(A) of the Endangered Species Act of 1973 requires a Recovery and Interstate Commerce Permit if propagation efforts will likely result in an endangered species being harassed, captured, harmed, possessed, or killed for "scientific purposes or to enhance the propagation or survival of the affected species." Examples of work that may require such a permit include, but are not limited to: abundance surveys, genetic research, relocation, capture and marking, and telemetric monitoring. Under certain circumstances, a permit also may be required to possess tissues and/or body parts of listed species. If you are doing habitat conservation work to help restore mussel populations in the United States, an Enhancement of Survival Permit also may be needed. In the United States, a federal permit does not exempt you from the state permitting process and different states have different permitting regulations. Countries outside the United States also have different permitting regulations. Contact the local natural resource agency to obtain the necessary permits before heading out to the field.

Be aware that the process of obtaining a new permit to work with endangered species in the United States can take a considerable time. Endangered species permit applications should be submitted well in advance of any proposed fieldwork. Having a detailed Propagation and Species Restoration Plan in place will help gather the required information for a permit.

1.5.12 *Do We Have a Plan for Collecting, Managing*
and Reporting Mussel Propagation Data?

Record-keeping is critical to the success of any mussel propagation program. Good records provide a means of evaluating the efficiency of the program over time and comparing results with other mussel propagation facilities. Data collection, however, is only one part of the story. It is also important to manage the data and compile it into a report (publication, annual report, etc.) that can be shared with other mussel propagation facilities. Data sharing between facilities can help coordinate propagation activities across the species entire range and help facilities learn from each other's successes and failures.

The kind of data collected is very important, but how the data is recorded is just as important. All records should be accurate, legible and thorough. At some point, someone will be trying to make decisions based on this data, so it is important to make sure it makes sense to the person compiling the data or writing the report.

When keeping records, one of the most important pieces of information to record is the date, including the day, month, and year the data was recorded. The four-digit numeric code should be used to represent the year portion of a calendar date (i.e. 2015) to avoid confusion. Without a clear date record, there is no frame of reference for the order of events or time elapsed since the event. Also include the name of the person recording the data in case questions arise later.

Key data that should be recorded throughout all steps of the mussel propagation process (i.e. host species collection, brood stock collection, host species infestation, juvenile culture, juvenile release, and monitoring) are listed below. Some sample data sheets have been included in the Appendix for reference.

1.5.12.1 Host Species Collection

During the collection or acquisition of the host species, it is important to record the following: the species of host collected, the number of each host species collected, collection method, collection date, collection location (GPS coordinates, state, county, major drainage, name of body of water, exact geographical location where the collection was made, and reference in miles and direction to a nearby town), names of the entire collection crew, and notes on the final disposition of the host species after propagation.

1.5.12.2 Brood Stock Collection

When collecting brood stock for mussel propagation, it is important to record the following: the species of mussel collected, the number of each species collected, collection method, collection date, collection location (GPS coordinates, state, county, major drainage, name of body of water, exact geographical location where the collection was made, and reference in miles and direction to a nearby town), water temperature, notes on habitat, names of the entire collection crew, brood stock holding location, holding system and length of time in holding system, brood stock release date, release location, and tag number.

1.5.12.3 Host Species Infestation and Juvenile Collection

For the host species infestation and juvenile collection process, it is important to record the following: infestation date, species of mussel, host species, the number of gravid female mussels and number of fish used, estimated fecundity, glochidia per fish in the infestation bath, condition of the glochidia, water temperature in the infestation bath, duration of the infestation, notes on fish condition including the gills and fins, host species holding location, holding system type and holding system water temperature, holding system maintenance (feeding, tank cleaning, etc.), host species survival and mortality during the larval attachment period, number of days the larvae are attached to the host, juvenile collection date, number of juveniles collected, and general condition of the juveniles coming off the host (actively pedal feeding, etc.).

1.5.12.4 Juvenile Culture

During the juvenile culture stage, it is important to record the following: culture location, type of culture system, culture system maintenance (water changes, cleaning, etc.), type of feed (wild water, pond water, cultured algae, etc.), feed concentration, water temperature, juvenile growth rates, and juvenile survival over time.

1.5.12.5   Juvenile Release

For juvenile release, it is important to record the following: release date, release location (GPS coordinates, state, county, major drainage, name of body of water, exact geographical location where the collection was made, and reference in miles and direction to a nearby town), names of the entire release crew, release method, conditions at the release site (water levels, water temperature, habitat, etc.), number of juveniles released, age of juveniles at the time of release, size range of juveniles at the time of release, condition of juveniles at the time of release and tag numbers for all juveniles released. If any permits were required for propagation and release, include the permit number on your data sheet.

1.5.12.6   Juvenile Monitoring

Monitoring will be discussed in more detail in Chapter 7; however, the following universal metrics should be recorded during the monitoring phase of the project: mussel bed area, mussel density, mussel size–frequency distribution, water temperature, and dissolved oxygen. It is also important to keep track of any tagged mussels collected during monitoring.

1.6   MANAGING AND PUBLISHING DATA

In any propagation and long-term monitoring program, large amounts of data will be generated so it is important to have a plan for managing the data. The key to any data management program is utility, including the ability to easily query data and summarize the results. Entire books and training courses exist on the subject of database management so this book will not go into great depth on this issue. The authors, however, would like to advocate for the development and maintenance of database standards for anyone working in mussel propagation. Database standards could help improve freshwater mussel recovery efforts by making data easier to understand, interpret, and share. The database management system should be able to organize data, create tables, query results, develop reports, and export data for use in other statistical applications.

Most importantly, all records should be compiled into a report (publication, annual report, etc.) that can be shared with other mussel propagation facilities, propagation biologists, and the mussel conservation community at large. The report should include detailed methods, data, and analysis for both propagation successes and failures. Transparency of methods and results is very important. Transparency will help other facilities build upon prior successes while preventing them from repeating previous failures or reinventing the wheel. If the time and staffing are not available to get the data in a report or publication, they should be archived and readily available. There are numerous publications that regularly publish papers on freshwater mussels and the Freshwater Mollusk Conservation Society publishes information on mussels and mussel propagation in both their journal (*Freshwater Mollusk Biology and Conservation*) and their newsletter (*Ellipsaria*).

2 Biology of Freshwater Mussels

Matthew A. Patterson

2.1 CLASSIFICATION

Freshwater mussels belong to the second largest phylum of animals in the world, the Mollusca. The phylum Mollusca includes approximately 90 000 species with a wide array of body sizes, shapes, and forms. Members of the phylum include octopus, squid, snails, slugs, oysters, scallops, chitons, and limpets. Oysters, scallops, clams, and the subject of this book, the freshwater mussels, belong in the class Bivalvia. Bivalve is a term used to describe the two shells (or valves) that enclose the soft tissues of the animal. The class Bivalvia includes approximately 15 000 species, the majority of which are found in the marine environment.

The freshwater bivalves comprise about 10% of all bivalve species and approximately 98% of those species can be found in two orders, the Veneroida and the Unionoida (Graf, 2013). The remaining species are found in marine bivalve orders that include a few freshwater representatives. The order Veneroida includes approximately 338 species and 20 families, including the fingernail clams (Sphaeriidae) and three species that are invasive to North America, the Asian clam (Cyrenidae = Corbiculidae) and the zebra mussel and quagga mussel (Dreissenidae). The order Unionoida includes approximately 855 species and 6 families, the Etheriidae (4 genera in Africa), Hyriidae (20 genera in Australasia and South America), Iridinidae (6 genera in Africa), Mycetopodidae (11 genera in South America), Margaritiferidae (9 genera in the Northern Hemisphere), and the Unionidae (167 genera in North America, Europe, Asia) (Graf, 2013). While both of these freshwater bivalve orders brood eggs inside the female ctenidia, the

Veneroida release either fully formed juveniles or free-swimming veliger larvae. The Unionoida, as we will discuss in great detail later, have a parasitic larval stage that in almost every case requires a host to complete metamorphosis to the juvenile stage.

The vast majority of the propagation and culture research to date has been focused on two families within the order Unionoida, the Unionidae and the Margaritiferidae. Consequently, this book will focus primarily on these two families; however, other families in the Order Unionoida will be discussed especially as it applies to their unique biology and life history.

Species in the family Margaritiferidae, commonly known as the "freshwater pearl mussels," have a very thick nacreous layer capable of producing high-quality pearls. The margaritiferids have a wide distribution including North America, Europe, northern Africa, the Middle East, and southern and eastern Asia (Smith, 2001). There has been debate over the years on the number of species and genera within this family (see a review by Smith, 2001) but it is clear that the margaritiferids make up a small percentage of the species in the Unionoida. Graf (2013) lists 13 species in the family or less than 2% of the known Unionoida.

By comparison, the family Unionidae includes approximately 681 species worldwide or 79% of the known Unionoida. Due to the high level of diversity in this family, it is helpful to discuss characteristics of the Unionidae at the subfamily and tribe level. Six subfamilies (Unioninae, Ambleminae, Gonideinae, Rectidentinae, Modellnaiinae, and Parreysiinae) are currently recognized worldwide (Graf and Cummings, 2006; Bieler et al., 2010; Whelan et al., 2011). Nearly 70% of the species in the family Unionidae fall in two subfamilies, the Unioninae (129 species) and the Ambleminae (334 species). Due to the high species diversity and focused propagation research efforts toward the Unioninae and Ambleminae, this book will focus primarily on propagation techniques developed for these two subfamilies.

Within the subfamily Unioninae there are two tribes, Unionini and Anodontini (Graf, 2013). The tribe Unionini includes 12 genera

(*Aculamprotula, Acuticosta, Arconaia, Cuneopsis, Diaurora, Inversiunio, Lanceolaria, Lepidodesma, Nodularia, Rhombuniopsis, Schistodesmus,* and *Unio*) that can be found in Europe, Asia and north Africa (Graf and Cummings, 2013). The tribe Anodontini includes 15 genera (*Alasmidonta, Anemina, Anodonta, Anodontoides, Arcidens, Cristaria, Lasmigona, Pegias, Pseudanodonta, Pyganodon, Simpsonaias, Simpsonella, Sinanodona, Strophitus,* and *Utterbackia*) that can be found in North America, Europe, and Asia.

Within the large subfamily Ambleminae there are four tribes, Amblemini, Lampsilini, Pleurobemini, and Quadrulini (Graf, 2013). The tribe Amblemini includes only one genus (*Amblema*), the tribe Lampsilini includes 25 genera (*Actinonaias, Arotonaias, Cyprogenia, Cyrtonaias, Delphinonaias, Disconaias, Dromus, Ellipsaria, Epioblasma, Friersonia, Glebula, Hamiota, Lampsilis, Lemiox, Leptodea, Ligumia, Medionidus, Obliquaria, Obovaria, Potamilus, Ptychobranchus, Toxolasma, Truncilla, Venustaconcha,* and *Villosa*), the tribe Pleurobemini includes 8 genera (*Pleurobema, Plethobasus, Hemistena, Fusconaia, Elliptoideus, Elliptio, Pleuronaia,* and *Uniomerus*), and the tribe Quadrulini includes 8 genera (*Quadrula, Amphinaias, Cyclonaias, Megalonaias, Plectomerus, Quincuncina, Theliderma,* and *Tritogonia*).

The freshwater mussel species within each tribe share common characteristics related to biology, life history, and propagation and culture techniques, so these tribes will be referenced throughout the book.

2.2 IDENTIFICATION

While accurate mussel identification is critical to any propagation program, this book will not address the intricacies of freshwater mussel identification. Because propagation requires working with live animals, mussels have to be identified using external shell characters. Identification based solely on external shell morphology can be problematic because environmental conditions, genetics, and other factors can lead to significant variation in shell shape, coloration, and thickness within a given species. Gently prying open the shell to look at internal, soft-body tissue characteristics like the color of the foot tissue can assist with identification. Consulting with local experts, published field guides, and museum collections is the best way to build mussel identification skills, but it can take years to master this skill. Shipping specimens to a local expert may provide the most accurate identification. Photographs of the side or lateral view of the shell as well as a close-up of the umbo (the oldest part of the shell) can be informative *in lieu* of shells in-hand. While this book is not intended to be a manual to freshwater mussel identification, the next section will cover a few of the key shell characters.

2.3   SHELL ANATOMY

Shell descriptions and dichotomous keys refer to the position of characters on the shell so it is important to get oriented to the anterior, posterior, dorsal, and ventral regions. In the Unionoida, the anterior of the shell is typically the area closest to the umbo (Figure 2.1). The umbo (also called the beak) is the oldest part of the shell and can be found along the dorsal margin, just anterior to the hinge ligament. The umbo and hinge ligament also are located in the dorsal region of the shell. The hinge ligament connects the two valves and serves to pull them apart when the adductor mussels are relaxed. The elevation of the umbo above the hinge line and position of the umbo on the shell can be important in identification. The sculpture of the umbo also can be useful in identification, but it can be eroded in older individuals. The posterior of the shell is the region furthest from the umbo. The posterior ridge and posterior slope are often referenced in keys and shell descriptions. The posterior ridge is a ridge running from the umbo to the posterior-ventral margin and the posterior slope is the area between the posterior ridge and the dorsal shell margin. The ventral region of the shell is where the two valves gape. Some keys and descriptions also reference the left valve or right valve. The left valve is the valve on the left when the dorsal edge is facing up and the anterior end is pointing away from you. The right valve is the

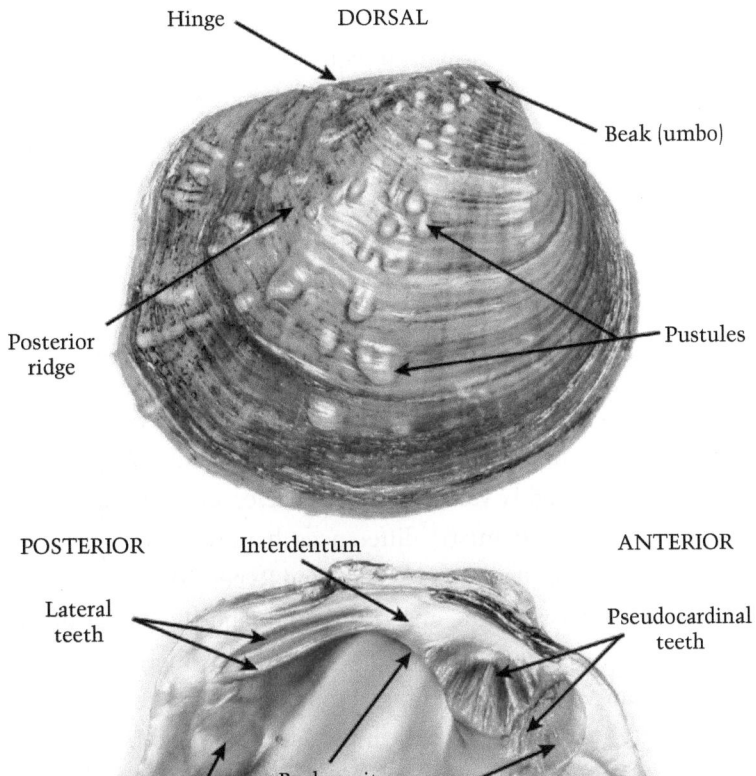

Hinge

DORSAL

Beak (umbo)

Posterior ridge

Pustules

POSTERIOR

Interdentum

ANTERIOR

Lateral teeth

Pseudocardinal teeth

Beak cavity

Muscle scars

Pallial line

VENTRAL

FIGURE 2.1 The left and right valve of the purple wartyback
(*Cyclonaias tuberculata*) showing some of the key characters for shell
identification.
Photo: Ryan Hagerty, USFWS. Graphics: Kristin Simanek, USFWS.
(A black-and-white version of this figure will appear in some formats.
For the color version, please refer to the plate section.)

valve on the right when the dorsal edge or hinge is facing up and the anterior end is pointing away from you.

Other diagnostic characters include the presence of a dorsal wing (a thin extension of the shell along the dorsal shell margin), a sulcus (a shallow depression anterior to the posterior slope), and a wide array of shell sculpture features including ridges, bumps, or spines. Shell sculpture can vary widely among and within species and can be present on the shell disk or posterior slope. Shell sculpture tends to be more common in larger rivers (Hornbach *et al.*, 2010).

Shell measurements (Figure 2.2) also can assist in identification. Shell length, height, and width (or thickness) can vary widely among species (35–280 mm in length and 1–20 mm in thickness). Length and width also can be highly variable within species as a result of latitudinal gradients, water chemistry differences (hard water vs. soft water), and food limitations (Bauer, 1992; Haag and Rypel, 2011). Shell length is the distance between the anterior margin and the posterior margin (typically measured parallel to the hinge line). Shell height is the greatest distance between the dorsal margin and the ventral margin and is usually perpendicular to the hinge line. Shell width is the greatest distance between the outer margins of the right and left valves. Shell growth in freshwater mussels is seasonal, leading to the presence of annual growth rings on the outer shell surface (Figure 2.3). These growth rings can only be used to estimate age because false growth rings can be laid down in response to stress.

Shell shape can be influenced by a number of factors. Ortmann's Law describes the tendency for shells to become more inflated with an increase in stream size (Ortmann, 1920). Shell inflation refers to the increase in shell width relative to height. Freshwater mussel species that inhabit lake environments are typically more inflated than stream species. Shell inflation can be important in identification, but much like shell sculpture, shell length, and shell thickness, it can vary widely among and within species. Some species of freshwater mussels also exhibit sexual dimorphism in shell shape. Sexual dimorphism is especially dramatic in the genus *Epioblasma* (Jones

FIGURE 2.2 Measuring freshwater mussel shells in the field using calipers.
Photo: Matthew Patterson, USFWS.

and Neves, 2010). In most cases, sexual dimorphism is caused by the swelling of the female gills during larval brooding (to be discussed later in this chapter).

The color of the outer shell layer (periostracum) may be diagnostic for some species, but can be highly variable. Color patterns may include background color, rays, and chevrons, but these patterns tend to become more difficult to see as the shell darkens with age.

Internal shell anatomy will be of limited use in identification, since propagation requires live mussels. With that said, dead shells

FIGURE 2.3 A left valve of the pimpleback (*Quadrula pustulosa*)
showing the prominent growth rings and shell sculpture.
Photo: Ryan Hagerty, USFWS.

along the bank can be helpful when trying to identify live specimens
in the field. Consequently, we will briefly discuss some of the commonly used internal shell characters here.

The hinge plate runs along the dorsal shell margin and includes
the pseudocardinal teeth and lateral teeth. The teeth on the left valve
interlock with the teeth on the right valve to help minimize twisting
or sheer stress on the shell when closed. The length, thickness, and
shape of both the lateral and pseudocardinal teeth may be important
in identification. The area between the lateral and pseudocardinal
teeth is called the interdentum, and the length and width of the interdentum may be important for identification. The umbo cavity or beak
cavity is the depression on the inside of the shell created by the raised
umbo. The cavity can be very deep in some species to nearly non-
existent in others. Muscle scars on the interior of the shell are attachment sites for the adductor muscles that close the shell. The nacre is
the shiny inner layer of the shell and nacre color may be helpful in
identification, but it can be highly variable within species.

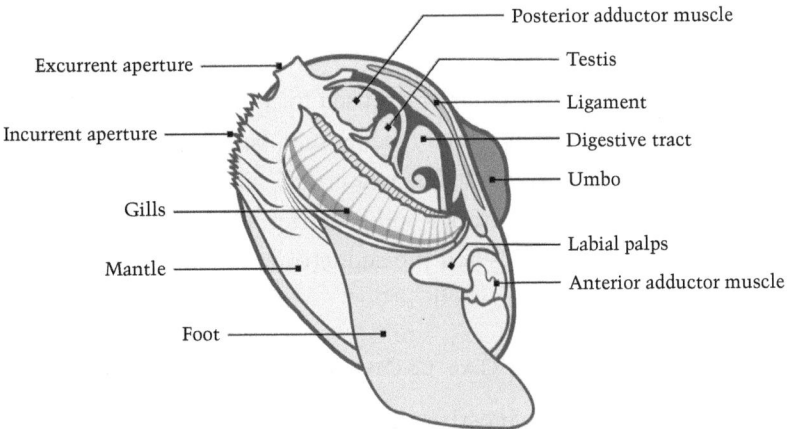

FIGURE 2.4 Soft tissue anatomy of a freshwater mussel.
Graphic: Kristen Simanek, USFWS (adapted from James Ford Bell
Museum).

2.4 SOFT TISSUE ANATOMY

A thorough understanding of the soft tissue anatomy of freshwater
mussels (Figure 2.4) is critical for propagation, especially the location
of the gills and larval brood pouches. In the Unionoida, the gills or
ctenidia are comprised of four demibranchs (inner and outer) with
two paired demibranchs on either side of the visceral mass (the visceral mass includes the gonad, digestive system, and other internal
organs). The ctenidia are used for gas exchange, suspension feeding,
and larval brooding. Due to their multifunctional role, the ctenidia
are typically very large structures that are relatively easy to see when
the shell is gently pried open. The ctenidia in the Unionidae are sub-
divided by septa to form vertical water tubes. The water tubes are
used to brood the larvae. The Margaritiferidae also brood the larvae
in the ctenidia, but the septa and water tubes are absent (Ortmann,
1912). Proper techniques for extracting larvae from the ctenidia are
discussed in Chapter 5.

The mantle is a thin layer of tissue that surrounds the soft tissues and functions to excrete the outer shell. The posterior mantle
margins merge to form incurrent and excurrent apertures. Water is

drawn into the shell through the incurrent aperture and exits through the excurrent aperture.

The foot is a muscular organ used for burrowing, locomotion, and feeding in juveniles. Pedal feeding may be an important means of obtaining food for juvenile freshwater mussels (Yeager *et al.*, 1994; Gatenby *et al.*, 1996). Anterior and posterior adductor muscles contract to close the shell. When the adductor muscles are relaxed, the hinge ligament opens the shell.

2.5   LOCOMOTION, GAS EXCHANGE, AND FEEDING

Freshwater mussels are mostly sessile, spending their life burrowed in the substrate. They use their muscular foot to move short distances and to burrow in the substrate. Newly metamorphosed juveniles can be seen using their foot to move around rather quickly in a Petri dish, presumably pedal sweeping to search for food. Juvenile freshwater mussels also use a byssus to attach to rocks and other hard substrates to help prevent downstream movement.

Freshwater mussels draw fresh water into the body cavity through the incurrent aperture. Cilia on the gills create water currents that direct fresh, oxygenated water into the water tubes through tiny holes (ostia or gill pores). Water then enters the suprabranchial cavity and passes out of the shell through the excurrent aperture. Water currents oxygenate the hemolymph (blood) and release carbon dioxide. The colorless hemolymph is circulated from the heart through the arteries to many small vessels throughout the body. The hemolymph returns to the gills where carbon dioxide is released and oxygen is added.

Suspended particles carried in the water currents are captured on the gills. In suboptimal conditions (including very high or very low food concentrations), bivalves have been shown to retract their feeding apertures, close the shell and/or reduce water pumping (Jorgensen, 1990). Particles are partially sorted and packaged into a mucilaginous string of material before being transported to the mouth for ingestion. Digestive enzymes and the crystalline style (a mortar and pestle-like organ) work together to digest and assimilate food particles inside the stomach. Newly metamorphosed juvenile mussels have small, rudimentary gills. Food particles collected through pedal feeding or deposit feeding have been shown to be important in obtaining food resources during this early life stage (Gatenby et al., 1996). Particles removed from suspension that are not ingested are packaged into "pseudofeces" and ejected through the ventral margin. Pseudofeces also can be released through the excurrent aperture. Excreta or fecal material are released through the excurrent aperture post-digestion.

Adult and juvenile freshwater mussels are very efficient filter feeders, capable of ingesting large amounts of suspended food particles over a 24-hour period. As a result, a continuous supply of food at the proper concentration is essential for growth and maintenance of condition in captivity. A continuous food supply also means the production of large amounts of waste (ammonia, CO_2, and fecal wastes). Removal of waste materials and uneaten feed will be important in the design and maintenance of aquaculture systems for both adult brood stock and juveniles.

2.6   REPRODUCTION

The gonad in freshwater mussels is located in the posteriodorsal portion of the visceral mass (Figure 2.4). Testes are whitish in color and ovaries are pinkish brown when observed macroscopically. In many species, both the males and females produce gametes year-round, but fully ripe gametes are only produced right before spawning. In other species, gametogenesis only occurs in the months leading up to spawning (Haggerty et al., 2005, 2011). Age of sexual maturity and first reproduction varies widely among species. Some fast-growing species may start producing eggs and sperm in the first year (Zale and Neves, 1982; Haag and Staton, 2003) while slower-growing species may not reach sexual maturity until much later. In the freshwater pearl mussel (Margaritifera margaritifera), for example, it can take up to 20 years to reach sexual maturity (Bauer, 1987). Seven mussel species in North America have been found to be hermaphroditic: the creek heelsplitter (*Lasmigona compressa*), the green floater (*Lasmigona subviridis*), the western pearlshell (*Margaritifera falcata*), the lilliput (*Toxolasma parvus*), the Savannah lilliput (*Toxolasma pullus*), the pondhorn (*Uniomerus tetralasmus*), and the paper pondshell (*Utterbackia imbecillis*) (van der Schalie, 1970; Hoeh *et al.*, 1995). Other species have been found to be occasionally hermaphroditic, but the rate of incidence is low and may occur in response to low population densities in the wild (Bauer, 1987).

Spawning occurs on a seasonal basis and seems to be based on water temperature (Hastie and Young, 2003). In the Unionidae, the timing of the spawn seems to be linked to the length of time female mussels brood the larvae in the gills. Mussel species that brood their larvae over the winter (long-term brooders) typically spawn in the late summer or autumn. Mussel species that spawn and release their larvae in the same calendar year (short-term brooders) typically spawn in the spring and early summer (although some species spawn in the fall). Multiple spawning periods per year have been observed for *Unio* (Engel, 1990; Fleischauer-Rossing, 1990; Hochwald, 1997), *Cumberlandia monodonta* (Gordon and Smith, 1990), and the Australian hyriids (Jones *et al.*, 1986).

Males release sperm into the water column as sperm balls or spermatozeugmata (Utterback, 1931; Edgar, 1965; Lynn, 1994). Spermatozeugmata are comprised of a thin, spherical membrane with thousands of attached spermatozoa. The sperm head is attached to the membrane with the flagellum pointing away from the membrane center. The adaptation of releasing spermatozeugmata, rarely found in marine bivalves, increases the longevity of the sperm cells in the freshwater environment and may allow for directional movement of the sperm balls (Barnhart and Roberts, 1997; Ishibashi *et al.*, 2000). While a single sperm cell may remain active for only a few minutes in fresh water, spermatozeugmata can remain active for up to 48 hours (Ishibashi *et al.*, 2000). This greatly increases the potential for long-distance sperm transport and fertilization. In fact, Ferguson *et al.* (2013) identified the father of three larvae located 16 km upstream of the brooding female in the plain pocketbook (*Lampsilis cardium*). If this sort of long-distance fertilization is widespread among the Unionidae, it is possible that reproduction in this group is not dependent on localized densities (Ferguson *et al.*, 2013).

The female mussel draws the spermatozeugmata into the shell through the incurrent aperture and fertilization is believed to occur inside the suprabranchial chamber. Following fertilization, the developing embryos and larvae are brooded inside the gills of the adult female and the portion of the gill tissues used for brooding is called the marsupium. Species in the family Margaritiferidae and the more primitive members of the subfamily Ambleminae are tetragenous, utilizing all four gills for brooding (Graf and Cummings, 2006). Most species within the Unionidae are ecto-branchous, utilizing only the outer demibranchs or some portion of the outer demibranchs for larval brooding. Members of the family Hyriidae are endobranchous, utilizing the inner demibranchs (Heard and Guckert, 1970). The amount of gill tissue dedicated to brooding also varies, with some species utilizing the entire length of the gill tissue and others utilizing only a portion of the gill (Williams *et al.*, 2008). Some evidence suggests the adult female provides nutrients for larval development during the brooding period (Schwartz and Dimock, 2001).

Parental care is very rare in the invertebrate world, but provides a distinct advantage in the freshwater environment. Most marine bivalves are broadcast spawners, releasing sperm and eggs into the water column for external fertilization. The resulting free-swimming larvae (veliger) eventually settle on the substrate to begin the adult stage. When marine bivalves invaded freshwater environments, the free-swimming larval stage likely was maladapted to the unidirectional currents that could send the larvae downstream to a saltwater environment. Parental care (larval brooding) in freshwater mussels enables the heavier weight and burrowing ability of the adult to help prevent the very small and lightweight larvae from getting washed downstream after fertilization (Cummings and Graf, 2010).

The duration of the brooding period is typically broken down into two categories: short-term brooders (tachytictic) and long-term brooders (bradytictic). In general, short-term brooders spawn early in the spring (some spawn in the fall) and brood the developing larvae only until they are fully mature and capable of parasitizing the host (Graf and Foighil, 2000). This entire process (spawning to metamorphosis) is typically completed over the course of four to five months, between early spring and the end of summer. Long-term brooders spawn in late summer and brood mature larvae in the gills throughout the winter. Release of the larvae and host infection for long-term brooders typically occurs the following spring; however, some species like the purple bean (*Villosa perpurpurea*) are more active in late winter (Brian Watson, personal communication). Variations in the brooding pattern have implications for propagation. In general, short-term brooders are more difficult to propagate because there is a shorter window of opportunity to collect mature larvae in the field. Collecting too early in the season can induce gravid females to release larvae before they are fully mature. Collecting too late in the season could lead to missing the brooding period and an entire year of propagation. Long-term brooders, on the other hand, have a much longer window of opportunity for collecting viable larvae in the field. Long-term brooders are also much less likely to prematurely release larvae in response to disturbance.

While there are exceptions to the following rule of thumb, long-term brooders in the family Unionidae tend to fall into two tribes, the Anodontini and Lampsilini. Short-term brooders tend to fall into the tribes Pleurobemini, Amblemini, and Quadrulini. Species in the family Margaritiferidae are more difficult to label as either long-term or short-term brooders. Some authors call them long-term brooders (Heard and Guckert, 1971; Davis and Fuller, 1981; Lydeard *et al.*, 1996), while others call them short-term brooders (Sterki, 1903; Ortmann, 1912), and some use terms like facultatively bradytictic (Graf and Foighil, 2000) and sequentially tachytictic (Heard, 1998). Graf and Foighil (2000) characterized the brooding pattern of margaritiferids as unknown. More research on the brooding cycle of the margaritiferids is clearly needed. From a propagation standpoint, it is helpful to know that margaritiferids behave like short-term brooders in that they have a tendency to release immature larvae or eggs when disturbed (Howard, 1915; Gordon and Smith, 1990). Understanding a mussel species' brooding cycle is critical for a successful propagation program.

There is some evidence that brooding females have reduced particle retention and particle transport rates along the gills compared to non-brooding mussels (Tankersley, 1996). Interestingly, two of the long-term brooding tribes (Anodontini and Lampsilini) have evolved separate strategies for dealing with reduced feeding rates during the extended brooding period. The Anodontini utilize the entire outer gill for brooding, but septa divide the water tubes into a central water tube that broods the larvae and a pair of secondary water tubes that continue to function in feeding and respiration (Ortmann, 1911; Richard et al., 1991; Tankersley and Dimock, 1993; Tankersley, 1996). The Lampsilini limit the effect of brooding on feeding and respiration by brooding the larvae in only one portion of the outer gill, typically the posterior-most portion of the outer gill (Cummings and Graf, 2010).

2.7   THE LARVAE

Two types of parasitic larvae can be found in the order Unionoida; glochidia and lasidia. The glochidium is found in the Unionidae, Margaritiferidae, and Hyriidae and is the most common larval stage in the Unionoida (Wachtler et al., 2001). The lasidium is found in the families Mycetopodidae, Iridinidae, and Etheriidae.

The term glochidium can be traced back to Rathke (1797) who erroneously believed that the small bivalves he found inside the gill tissues were parasites of the adult mussel. Rathke gave them the Latin name, *Glochidium parasiticum*. Although his theory was later disproved (Carus, 1832; Leydig, 1866), the name glochidium stuck.

The number of glochidia per female can range from several thousand to several million depending on both the size of the glochidia and the size of the female (Wachtler et al., 2001). Mature glochidia have two calcified shells held together by a single adductor

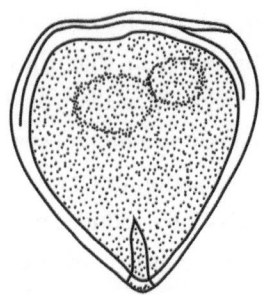 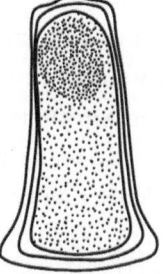 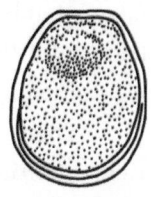

FIGURE 2.5 The three primary types of glochidia in the North American Unionidae. Subtriangular, hooked glochidia (left), axe-head glochidia (center), and semi-elliptical, hookless glochidia (right).
Graphic: Kristin Simanek, USFWS (adapted from Surber, 1912).

mussel, sensory hairs, and a larval thread (in some species), but lack most other anatomical features of the adult (Wachtler *et al.*, 2001). Three basic morphological groups of glochidia (Figure 2.5) are recognized: (1) subtriangular, hooked larvae, (2) axe-head larvae, and (3) semi-elliptical, hookless larvae (Kat, 1984). Hooked glochidia are found only in the subfamily Unioninae (i.e. *Pyganodon, Lasmigona, Alasmidonta*) and the Hyriidae of Australia and South America (Graf and Cummings, 2006; Cummings and Graf, 2010). They are large in size (200–380 microns in length) and have large hooks or spines on the ventral shell margin. The axe-head glochidia are found only in one genus of freshwater mussel (*Potamilus*). A very large height to length ratio (length = 218 microns; height = 379 microns) and expansion of the shell along the ventral margin gives this small group a very unique shape. Like the hooked glochidia, small hooks or spines are found on the ventral shell margin in the axe-head form. Hookless glochidia, found in the Ambleminae and Margaritiferidae, are smaller (60–277 microns in length) and have no hooks or spines on the shell. Depending on the species and water temperature, glochidia may survive between 2 and 14 days outside the marsupium (Fisher and Dimock, 2000; Zimmerman and Neves, 2002).

The second type of parasitic larva in the order Unionoida is the lasidium. Lasidia also are brooded within the ctendia of the adult female and range in size from 150–200 microns (Wacthler *et al.*, 2001). Unlike glochidia, lasidia are enclosed within a single shell that is not calcified (Mackie, 1984) and have an extraordinarily long larval thread (Wachtler *et al.*, 2001). In *Mutela bourguignati*, the larval thread can be 15 mm in length or over 70× the length of the shell (Fryer, 1961). The long larval thread is thought to increase buoyancy and increase the chance of contact with the fish host (Fryer, 1961). Once in contact with the host, lasidia attach with the use of posterior hooks and ciliated lobes (Barnhart *et al.*, 2008).

2.8   STRATEGIES FOR HOST INFECTION

The larvae of a few species in the order Unionoida (*Utterbackia imbecilis, Strophitus undulatus, Obliquaria reflexa, Lasmigona sub-viridis, Grandidieria burtoni*) and several mussel species in South America appear to be capable of completing metamorphosis inside the gills of the adult female (Parodiz and Bonetto, 1963; Mackie, 1984; Kondo, 1990; Mansur and Campos-Velho, 1990; Barfield and Watters, 1998; Heard, 1998; Lellis and King, 1998; Dickinson and Sietman, 2008). For mussel species that do not exhibit direct development to the juvenile stage, the mature larvae must parasitize a host to complete metamorphosis. A parasitic larval stage is unique among the bivalves (Cummings and Graf, 2010) and most freshwater mussel species utilize freshwater fish as a host. At least one species, the salamander mussel, (*Simpsonaias ambigua*), utilizes the mudpuppy (*Necturus maculosus*) as a host. By parasitizing a mobile host, the larvae can be protected from downstream movement and even move upstream against the current and disperse to new areas.

Some mussel species are host generalists (using a wide array of host species), while others are considered host specialists (using relatively few host species or even a single species). Freshwater mussels are mostly sessile organisms, so attaching the larvae to a highly mobile host like a fish can be problematic. For mussel species that are host specialists, the problem is even more complex because they must ensure the larvae attach to the appropriate, compatible host.

The adaptations that have evolved in the freshwater mussels to increase the probability of larval attachment to a host are what make freshwater mussels one of the most fascinating groups of organisms in the world.

Barnhart *et al.* (2008) and Haag (2012) provide excellent reviews of these intriguing host infection strategies for both the Unionidae and the Margaritiferidae, but we will provide a brief overview here.

2.8.1   *Broadcasting*

Broadcasters release larvae into the water column. Because they rely largely on chance encounters with the host, broadcasters tend to be host generalists. Most of the freshwater mussels in the tribe Anodontini and the family Margaritiferidae are broadcasters, suggesting this is the more primitive strategy (Haag, 2012). Haag (2012) further breaks down broadcasters into two major substrategies: passive entanglement and broadcast of free larvae. In passive entanglement, adult females release larvae in mucus strings or the larvae produce a "larval thread." Both structures serve to suspend the larvae in the water column and entangle the host, increasing the chances of attachment (Lefevre and Curtis, 1912; Wood, 1974; Aldridge and McIvor, 2003). Fryer (1961) concluded that the long larval threads of the lasidia of *Mutela bourguignati* serve to increase contact with the host through entanglement.

Species that broadcast free larvae into the water tend to be host specialists. In this case, the odds of attachment would seem almost insurmountable, but these species have some adaptations to help overcome the odds. Populations of the freshwater pearl mussel (*Margaritifera margaritifera*), for example, combine high fecundity with synchronous release to flood the water column with huge numbers of larvae at a time when their migratory salmonid host is locally abundant (Young and Williams, 1984; Barnhart *et al.*, 2008). Despite low fecundity, the salamander mussel (*Simpsonaias ambigua*) overcomes the odds by sharing tight quarters with their host. Both the adult mussels and their host, the mudpuppy (*Necturus maculosus*), live in tight spaces under rocks, increasing the chances of attachment (Howard, 1915).

2.8.2 Conglutinates

Some freshwater mussel species release their larvae in small packages called conglutinates (Lefevre and Curtis, 1912). Conglutinates, formed inside the water tubes of the adult female, increase the probability of larval attachment by resembling preferred food items of the host. In the genus *Ptychobranchus*, for example, conglutinates can mimic chironomid larvae (Hartfield and Hartfield, 1996), the larvae and pupae of blackflies (Barnhart *et al.*, 2008), and fish fry (Watters, 1999). Conglutinates often contain undeveloped or unfertilized eggs that provide both structure and color (Barnhart, 1997). Unfertilized eggs represent a loss of reproductive output, but this loss is balanced with an increase in the chance of host attachment (Barnhart *et al.*, 2008). Haag (2012) further divides conglutinate releasers into the following categories: (1) pelagic, (2) demersal, (3) mucoid, and (4) superconglutinates.

Pelagic conglutinates can be found in a few genera in the tribe Pleurobemini (i.e., *Fusconaia*, *Pleurobema*, and *Pleuronaia*). They are small conglutinates that drift in the water column. Pelagic conglutinates come in a wide array of shapes and colors (Figure 2.6). Pigmentation pattern and structure arise primarily from a high percentage of undeveloped or unfertilized structural eggs (Haag and Warren, 2003; Barnhart *et al.*, 2008). The freshwater mussel species that produce pelagic conglutinates tend to be host specialists, attaching to minnows that feed on particulates floating in the water column (e.g. *Cyprinella* and *Notropis*).

Demersal conglutinates can be found in a few genera in the tribe Lampsilini (i.e., *Cyprogenia*, *Dromus*, *Obliquaria*, and *Ptychobranchus*). In contrast to the pelagic variety, demersal conglutinates are found along the bottom of the stream (Figure 2.7). The mussel species that produce demersal conglutinates also tend to be host specialists, focusing on benthic-dwelling fish species like darters or sculpins. Conglutinates in this group mimic a variety of benthic food

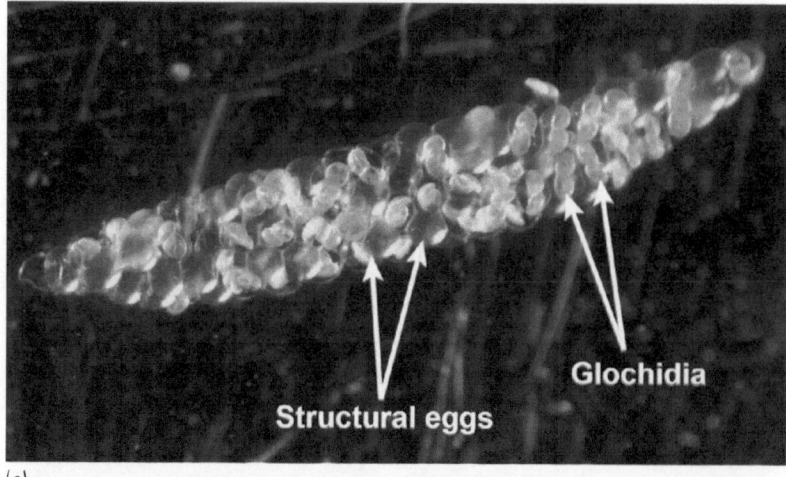

(a)

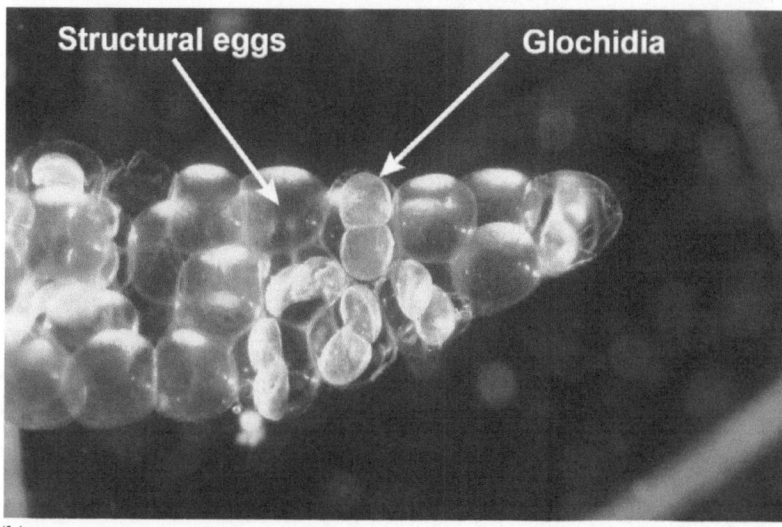

(b)

FIGURE 2.6 (a) The pelagic conglutinate of the Wabash pigtoe (*Fusconaia flava*). The reddish-orange color is provided by the structural eggs. (b) A close-up view of pelagic conglutinate of the Wabash pigtoe (*Fusconaia flava*) showing the bivalved glochidia and the reddish-orange structural eggs.

Photos: Bernard Sietman, Minnesota Department of Natural Resources. (A black-and-white version of this figure will appear in some formats. For the color version, please refer to the plate section.)

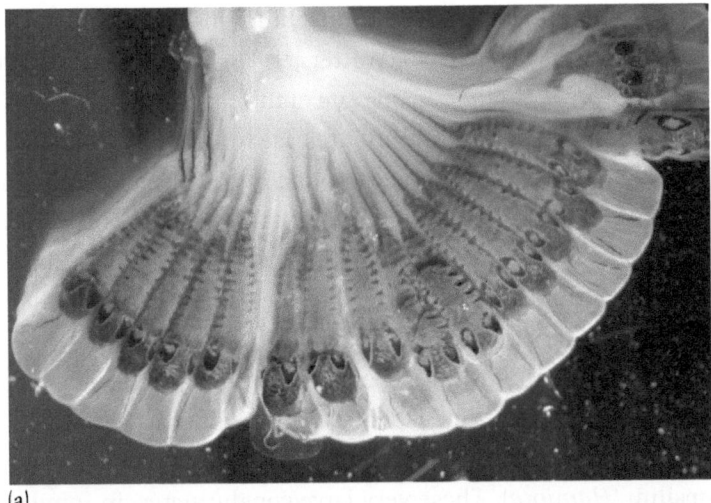

(a)

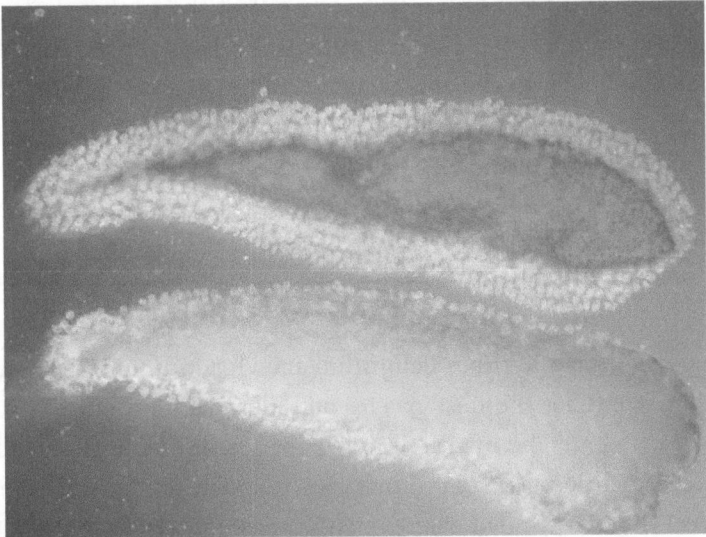

(b)

FIGURE 2.7 (a) The demersal conglutinate of the fluted kidneyshell (*Ptychobranchus subtentum*). The conglutinate closely resembles the pupal stage of a black fly. (b) The demersal conglutinate of the Dromedary pearlymussel (*Dromus dromas*). The mature glochidia can be seen around the edges of the conglutinate next to the structural eggs that closely resemble blood worms or fish eggs.

Photos: Rachel Mair, USFWS. (A black-and-white version of this figure will appear in some formats. For the color version, please refer to the plate section.)

resources, such as leaches, worms, blackflies, fish eggs, or larval fish (Hartfield and Hartfield, 1996; Watters, 1999; Barnhart *et al.*, 2008). In the genus *Ptychobranchus*, one end of the conglutinate is adhesive, allowing it to hold on to rocks after release from the adult female.

Mucoid conglutinates can be found in a few genera in the tribe Quadrulini (i.e., *Cyclonaias* and *Quadrula*). Mucoid conglutinates also are associated with benthic habitats; however, they lack the clear structure found in the demersal conglutinates. Instead they are a loose mass of larvae and mucus that may resemble worms or insect larvae. The freshwater mussel species that produce mucoid conglutinates tend to be host specialists that utilize catfish as a host.

Superconglutinates can be found in only one genus in the tribe Lampsilini (*Hamiota*). These very large conglutinates, in some cases containing all of the larvae from one gill of the adult female, are released into the water column as part of a large mucus tube that remains attached to the adult female and undulates with the river current (much like a fishing line and lure). All of the species that utilize superconglutinates are host specialists that utilize black bass as their host. The bass are attracted to the mucus tube, which strongly resembles wounded or distressed prey.

2.8.3 *Mantle Lures*

Some species in the tribes Lampsilini and Quadrulini utilize a mantle lure to attract the host species and increase the chances of larval attachment. The mantle lure is a modified portion of the adult female's mantle tissue that oftentimes resembles the preferred prey item of the host species. The lure also can be undulated in the water column to mimic the prey's movement pattern. In some cases, special musculature allows the female to position the marsupium in close proximity to the mantle lure. When the host strikes the lure, the marsupium is ruptured and the larvae are released (Barnhart *et al.*, 2008). Haag (2012) grouped mantle lures into three categories (large lures, cryptic lures, and mantle magazines) based on general structure and the host species targeted.

Large lures are found in only a few genera in the tribe Lampsilini (i.e. *Hamiota, Lampsilis, Ligumia,* and *Villosa*). The lures typically resemble small prey items like small fishes, crayfish, and aquatic insects (Figure 2.8a) and are found in freshwater mussel species with large, predatory hosts. The mantle lure of the plain pocketbook (*Lampsilis cardium*) resembles a small minnow, including pronounced eye spots, fins, and lateral line (Figure 2.8b). In this species, the mantle lure can be undulated in the water, mimicking the swimming motion of a small minnow. The protrusion between the two halves of the mantle lure is the marsupium charged with glochidia. When the host species (in this case a smallmouth bass, *Micropterus dolomieu*) attacks the mantle lure, the water tubes are ruptured and the larvae released.

Cryptic lures are found only in a few genera in the tribe Lampsilini (i.e. *Epioblasma, Lemiox, Ligumia, Medionidus, Obovaria, Toxolasma, Venustaconcha,* and *Villosa*). Smaller than the large lures, cryptic lures (Figure 2.9) attract smaller host species, including darters and sculpin. As with the large lures, many cryptic lures mimic food items of the host species (small snails, aquatic insects, fish eggs, etc.). One genus, *Epioblasma*, uses a cryptic lure to trap the host species, usually by the head, inside the shell of the female (Mulcrone, 2004; Jones *et al.*, 2006b; Barnhart *et al.*, 2008). Once the host species is trapped, larvae are released and attach. Teeth along the margin of the shell aid in holding the host species in place while the larvae attach. In contrast to large lure species that require the water tubes to be broken open by the host, the genus *Epioblasma* actively releases larvae from the water tubes while the host species is trapped.

Mantle magazines are found only in a few genera in the tribe Quadrilini (i.e., *Cyclonaias* and *Quadrula*). This extension of the mantle tissue serves as a storage area for larvae and conglutinates, as well as an attractant for host species (Figure 2.10). When the host species attacks, stored larvae can quickly be released for attachment (Barnhart *et al.*, 2008). The mapleleaf (*Quadrula quadrula*) and pimpleback (*Quadrula pustulosa*) groups, along with the pistolgrip

(a)

(b)

FIGURE 2.8 (a) The mantle lure of the wavy-rayed lampmussel (*Lamspilis fasciola*). (b) The mantle lure of the plain pocketbook (*Lampsilis cardium*) with fake eye spots, fins, and lateral line. The tissue bulging between the two sides of the mantle lure is the marsupial gill with brooding larvae.

Photos: (a) Rachel Mair, USFWS. (b) Ryan Hagerty, USFWS. (A black-and-white version of this figure will appear in some formats. For the color version, please refer to the plate section.)

FIGURE 1.3 Fresh-dead shells of a male (top) and female (bottom) purple catspaw (*Epioblasma obliquata obliquata*). Photo: Angela Boyer, USFWS.

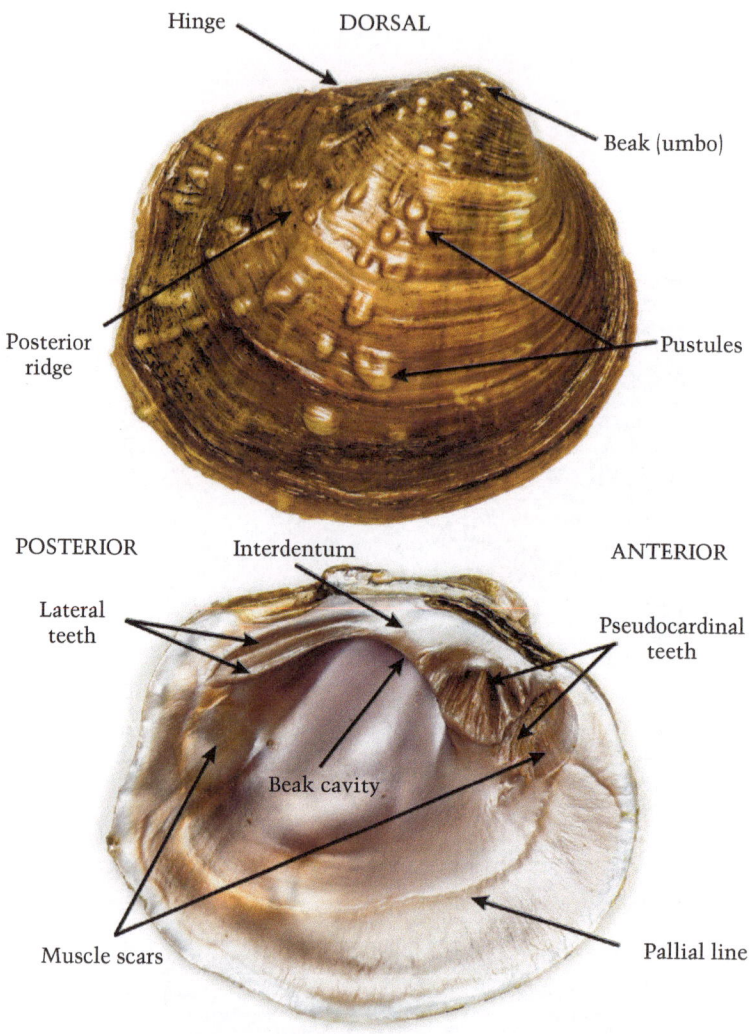

FIGURE 2.1 The left and right valve of the purple wartyback (*Cyclonaias tuberculata*) showing some of the key characters for shell identification.
Photo: Ryan Hagerty, USFWS. Graphics: Kristin Simanek, USFWS.

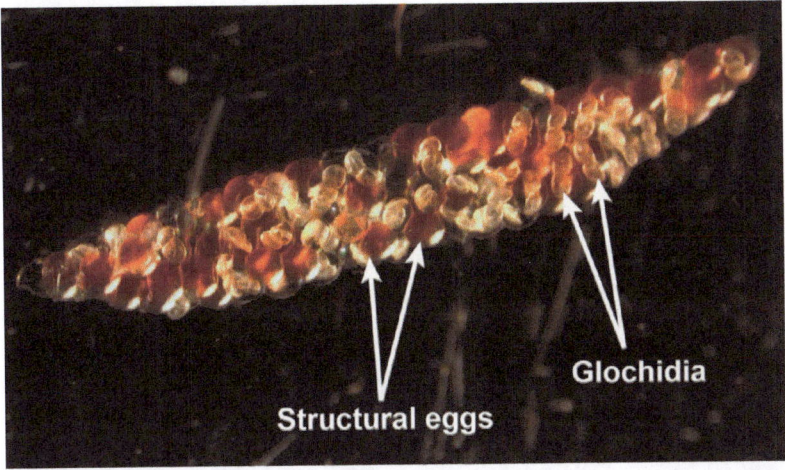

(a)

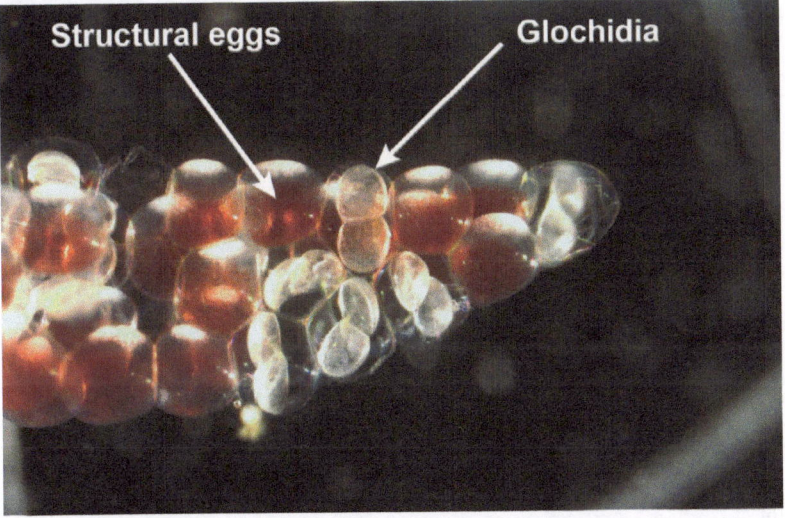

(b)

FIGURE 2.6 (a) The pelagic conglutinate of the Wabash pigtoe (*Fusconaia flava*). The reddish-orange color is provided by the structural eggs. (b) A close-up view of pelagic conglutinate of the Wabash pigtoe (*Fusconaia flava*) showing the bivalved glochidia and the reddish-orange structural eggs.

Photos: Bernard Sietman, Minnesota Department of Natural Resources.

(a)

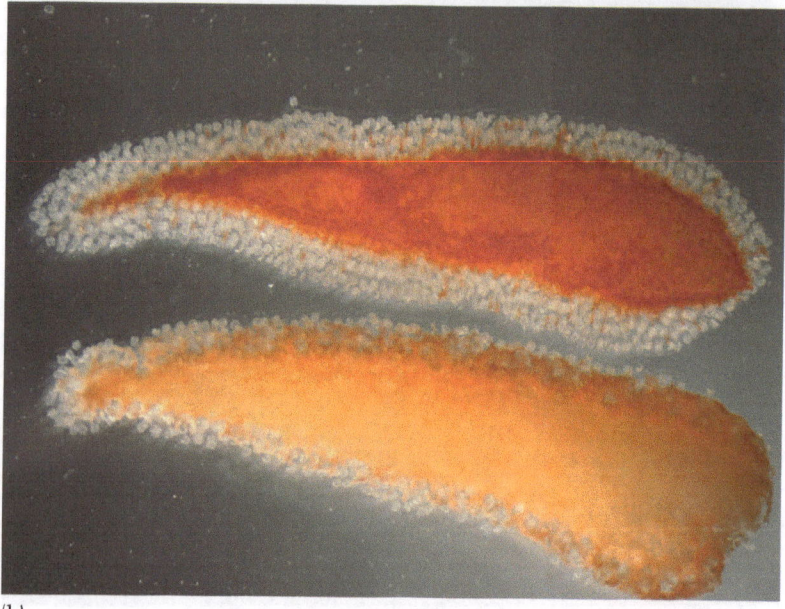

(b)

FIGURE 2.7 (a) The demersal conglutinate of the fluted kidneyshell
(*Ptychobranchus subtentum*). The conglutinate closely resembles
the pupal stage of a black fly. (b) The demersal conglutinate of the
Dromedary pearlymussel (*Dromus dromas*). The mature glochidia can
be seen around the edges of the conglutinate next to the structural eggs
that closely resemble blood worms or fish eggs.
Photos: Rachel Mair, USFWS.

(a)

(b)

FIGURE 2.8 (a) The mantle lure of the wavy-rayed lampmussel
(*Lamspilis fasciola*). (b) The mantle lure of the plain pocketbook
(*Lampsilis cardium*) with fake eye spots, fins, and lateral line. The
tissue bulging between the two sides of the mantle lure is the marsupial
gill with brooding larvae.
Photos: (a) Rachel Mair, USFWS. (b) Ryan Hagerty, USFWS.

(b)

FIGURE 2.9 (b) The microlure of the Cumberlandian combshell
(*Epioblasma brevidens*) closely resembles fish eggs.
Photo: Rachel Mair, USFWS.

FIGURE 2.11 Glochidia of the plain pocketbook (*Lampsilis cardium*)
attached to the gill filaments of a largemouth bass (*Micropterus
salmoides*).
Photo: Molly Webb, USFWS.

(a)

FIGURE 3.7 (a) Labeling all quarantine equipment with colored tape can help prevent transfer of that equipment to clean areas of the propagation facility.
Photo: Matthew Patterson, USFWS.

FIGURE 4.1 Collecting a buccal swab from a sheepnose (*Plethobasus cyphyus*) for genetic analysis.
Photo: Bryan Simmons, USFWS.

(b)

FIGURE 4.3 (b) Freshwater mussels being held in collecting bags in the river. The mussels are laid flat on the bottom of the river in a single layer to maintain condition.
Photo: Matthew Patterson, USFWS.

Non-Gravid Gill

(a)

FIGURE 4.6 (a) The Eastern lampmussel (*Lampsilis radiata*) gently pried open to reveal the gills and check for gravidity. The flattened gills in this female indicate it is not gravid.
Photo: Matthew Patterson, USFWS. Graphics: Kristin Simanek, USFWS.

(b)

FIGURE 4.6 (b) The inflated gills of a female plain pocketbook (*Lampsilis cardium*), indicating this female mussel is gravid and brooding glochidia.
Photo: Ryan Hagerty, USFWS. Graphics: Kristin Simanek, USFWS.

(d)

FIGURE 5.11 (d) Darters being held in a 10 L Aquatic Habitat tank for infestation. Enough water is added to just cover the backs of the fish host.
Photo: Nathan Eckert, USFWS.

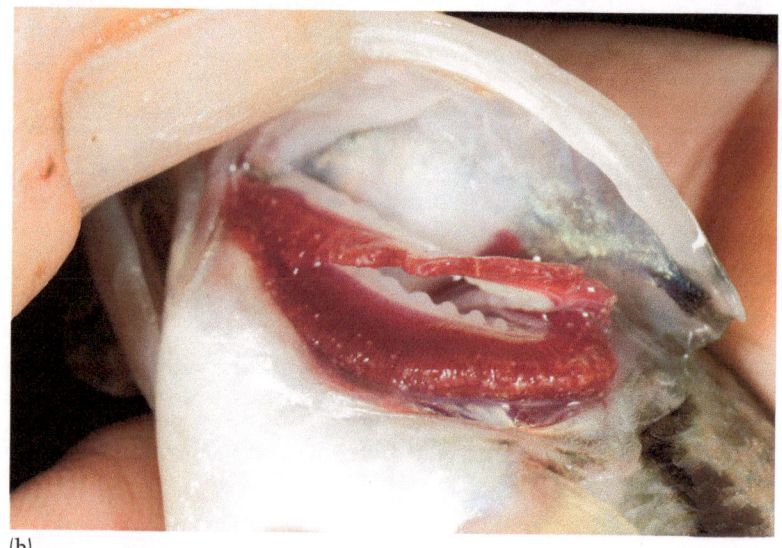

(b)

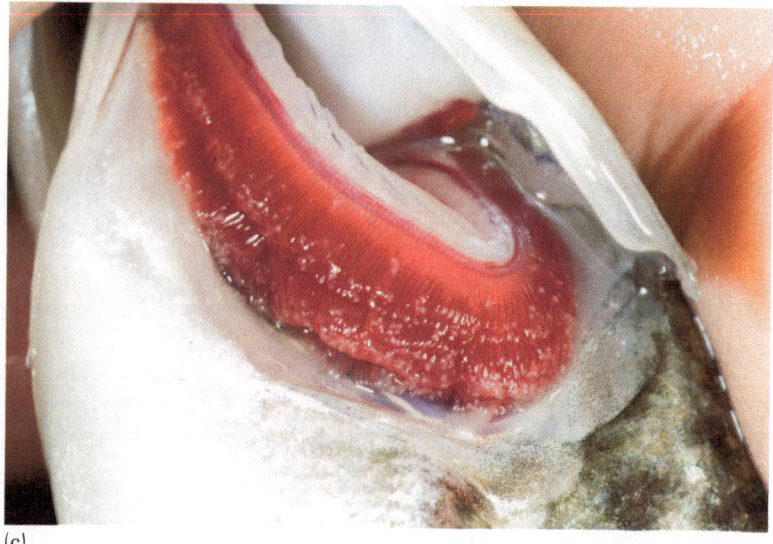

(c)

FIGURE 5.13 (b) The gills of the fish host after 5 minutes in the infestation bath. The small white specks on the gill are attached glochidia. (c) The gills of the fish host after 30 minutes in the infestation bath. The number of attached glochidia has increased significantly, forming a white ring around the edge of the gill. Photos: Ryan Hagerty, USFWS.

FIGURE 5.16 A customized multi-tank rack system at the Virginia Department of Game and Inland Fisheries' Aquatic Wildlife Conservation Center outfitted with 10 L tanks for larger-scale production. Photo: Matthew Patterson, USFWS.

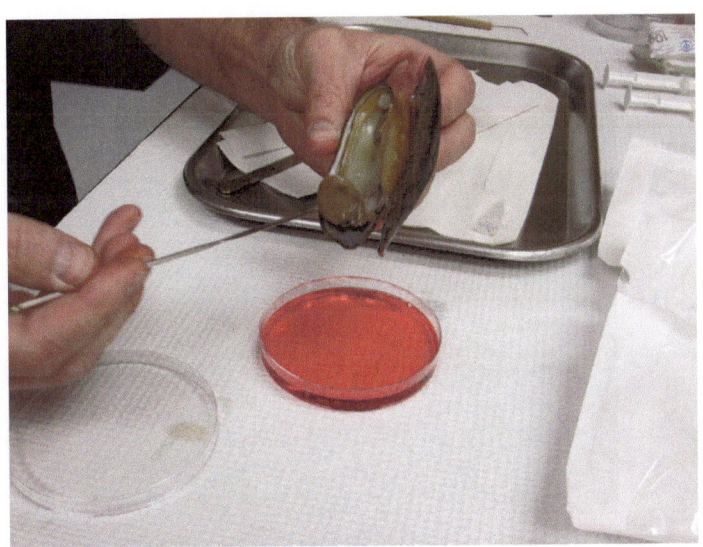

(a)

FIGURE 5.18 (a) Excision of the demibranchs for *in vitro* culture at the Warm Springs National Fish Hatchery (Warm Springs, Georgia). This lethal extraction method should not be used for imperiled species. Photo: Jaclyn Zelko, USFWS.

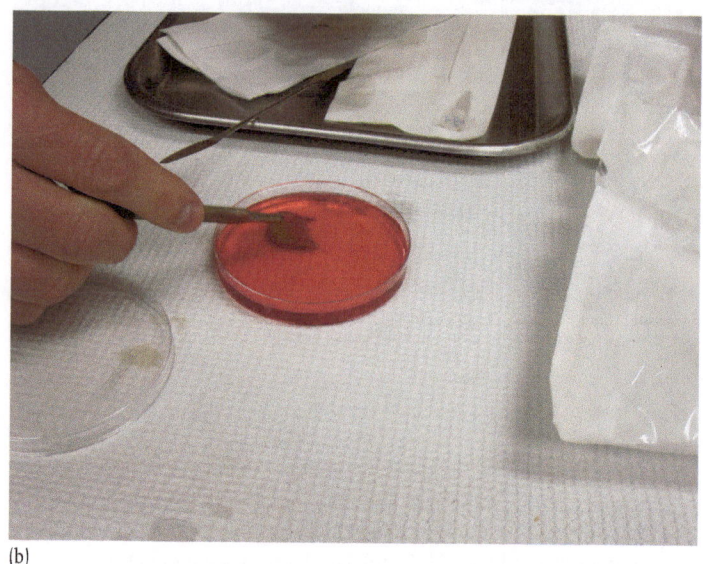

(b)

FIGURE 5.18 (b) Excised gills being placed in the culture media to complete metamorphosis.
Photo: Jaclyn Zelko, USFWS.

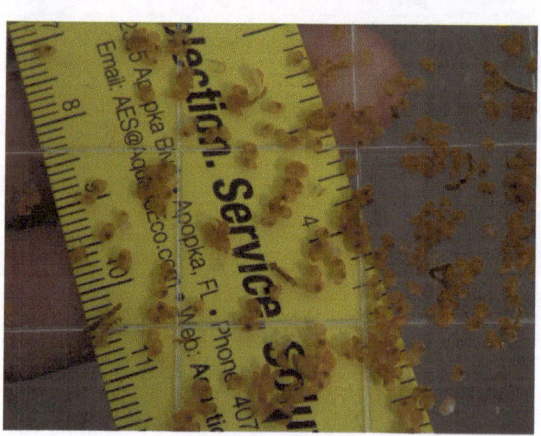

FIGURE 5.21 Juvenile mussels in a gridline Petri dish. The grid assists with the enumeration process.
Photo: Rachel Mair, USFWS.

FIGURE 6.1 Subadult mucket (*Actinonaias ligamentina*) tagged for release.
Photo: Matthew Patterson, USFWS.

(d)

FIGURE 6.5 (d) A single rearing pan with juvenile freshwater mussels feeding at the surface.
Photo: Rachel Mair, USFWS.

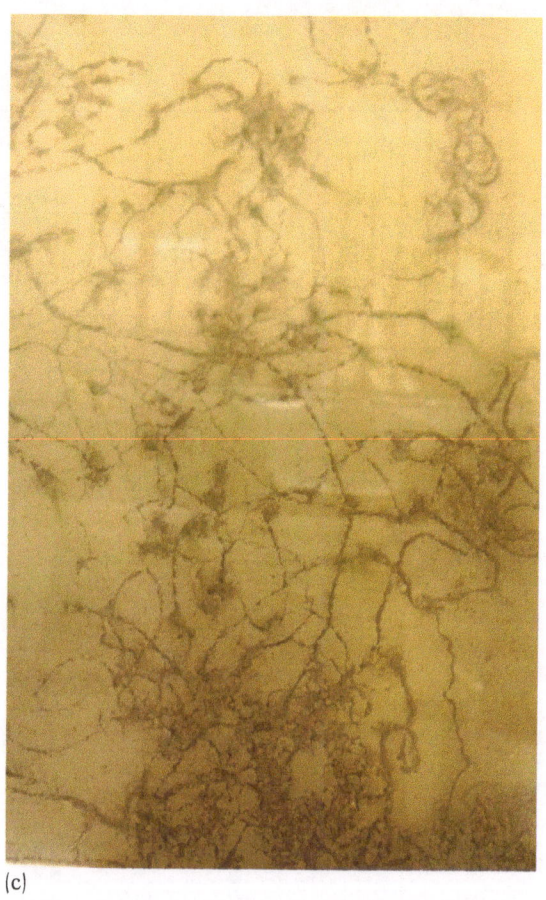

(c)

FIGURE 6.6 (c) Juvenile mussels making trails in the substrate in a
Hruska box.
Photo: Beth Glidewell, Missouri State University.

(e)

FIGURE 6.7 (e) Juvenile Alabama lampmussel (*Lampsilis virescens*) inside the partial flow-through bucket system culture chamber. Photo: Paul Johnson, Alabama Aquatic Biodiversity Center.

(e)

FIGURE 6.10 (e) Juvenile pink mucket (*Lampsilis abrupta*) being reared in the raceway upweller at the Kansas City Zoo. Photo: Chris Barnhart, Missouri State University.

(d)

FIGURE 6.11 (d) Outdoor mussel culture cages on floating racks at Genoa National Fish Hatchery (Genoa, WI). The white flotation buoy in the upper right hand corner of the photo is used to keep the cages off the bottom of the pond.
Photo: Ryan Hagerty, USFWS.

(f)

FIGURE 6.12 (f) Subadult eastern pondmussel (*Ligumia nasuta*) and yellow lampmussel (*Lampsilis cariosa*) inside a floating basket at the Virginia Fisheries and Aquatic Wildlife Center at Harrison Lake National Fish Hatchery (Charles City, VA).
Photo: Matthew Patterson, USFWS.

FIGURE 6.18 Four-month-old alewife floater (*Anodonta implicata*) cultured in floating baskets on wild pond water by the Virginia Fisheries and Aquatic Wildlife Center at Harrison Lake National Fish Hatchery. Photo: Rachel Mair, USFWS.

(c)

FIGURE 6.14 (c) The inside of a SUPSY culture chamber with juvenile mussels resting on the mesh screen. Photo: Ryan Hagerty, USFWS.

(a)

(b)

FIGURE 7.1 (a) A SCUBA diver releasing tagged, subadult northern
riffleshell (*Epioblasma torulosa rangiana*) to the Allegheny River.
(b) Subadult fatmucket (*Lampsilis siliquoidea*) tagged with Hallprint
tags and transported to the release site in a mesh bag.
Photos: (a) Janet Butler, USFWS. (b) Rachel Mair, USFWS.

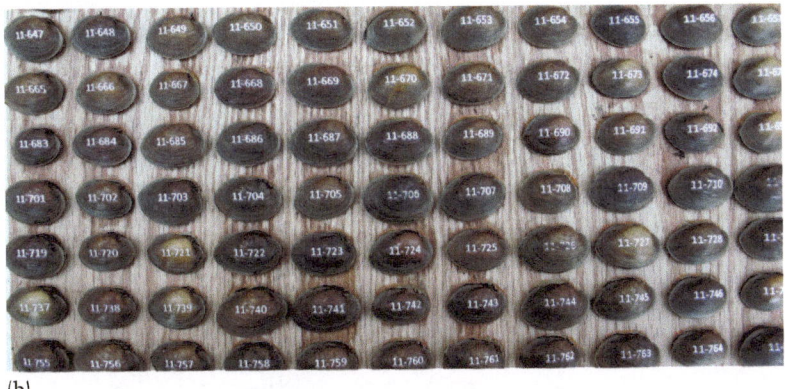

(b)

FIGURE 7.3 (b) A large cohort of sub-adult pink mucket (*Lampsilis abrupta*) tagged with a laser engraver.
Photo: Chris Barnhart, Missouri State University.

(a)

FIGURE 7.4 (a) Higgins eye pearly mussel (*Lampsilis higginsi*) tagged with both a black glue dot and a Hallprint tag.
Photo: Gary Wege, USFWS.

(c)

FIGURE 7.4 (c) A northern riffleshell (*Epioblasma torulosa rangiana*) tagged with a purple Hallprint tag that is starting to fade to blue in the river.

Photo: Jeremy Tiemann, Illinois Natural History Survey.

(a)

FIGURE 7.5 (a) PIT tags coated in white epoxy in combination with Hallprint tags can be used to monitor imperiled species like the northern riffleshell (*Epioblasma torulosa rangiana*).

Photo: (a) Rachel Mair, USFWS.

(d)

FIGURE 7.5 (d) A PIT tag reader being operated by a SCUBA diver to detect tagged mussels in a large river.
Photo: (d) Janet Clayton, West Virginia Division of Natural Resources.

(b)

FIGURE 7.6 (b) Close-up of mussels loaded in the plastic dish racks.
Photo: Janet Clayton, West Virginia Division of Natural Resources.

FIGURE 7.9 (a) Quantitative sampling for freshwater mussels. The substrate is excavated from the PVC quadrat and sieved to search for adult and juvenile mussels. (b) Completing a timed sample of a long-term monitoring grid on the Cacapon River in West Virginia. The orange flags indicate the position of a freshwater mussel in the PVC grid. Photos: Matthew Patterson, USFWS.

(a)

FIGURE 8.6 (a) Algae culture slants, flasks, and carboys in the clean lab
at the White Sulphur Springs National Fish Hatchery.
Photo: Matthew Patterson, USFWS.

(b)

FIGURE 8.6 (b) Bulk algae cultures growing in a Biofence in the greenhouse at the White Sulphur Springs National Fish Hatchery. Photo: Matthew Patterson, USFWS.

(a)

(b)

FIGURE 2.9 (a) Female oyster mussel (*Epioblasma capsaeformis*) gaping at the substrate surface to attract a host. The iridescent mantle tissue and undulating microlure entice a fish to enter the shell where it is captured with the help of the teeth along the shell margin. (b) The microlure of the Cumberlandian combshell (*Epioblasma brevidens*) closely resembles fish eggs. (c) Display of the northern riffleshell (*Epioblasma torulosa rangiana*).

Photos: Rachel Mair, USFWS. (A black-and-white version of part (b) of this figure will appear in some formats. For the color version, please refer to the plate section.)

(c)

FIGURE 2.9 (*cont.*)

(*Tritogonia verrucosa*) and the purple wartyback (*Cyclonaias tuber-culata*) are catfish specialists, whereas the monkeyface (*Quadrula metanevra*) group specialize on minnows (Barnhart *et al.*, 2008; Hove *et al.*, 2011; Seitman *et al.*, 2012).

2.8.4 Freshwater Drum

Several species of large-river freshwater mussels are specialists on freshwater drum (*Aplodinotus grunniens*), including *Ellipsaria*, *Leptodea*, *Potamilus*, and *Truncilla*. Some unique strategies have evolved in these species for infesting freshwater drum which consume mollusks and other food resources by feeding along the river bottom. More research is needed on the infestation strategies in this group, but some evidence suggests that these strategies may include: (1) gravid females broadcasting larvae along the river bottom where the drum are feeding, (2) gravid females gaping widely at the surface to attract the drum which extract the larvae via suction, and (3) drum ingesting gravid females of thin-shelled species (Coker *et al.*, 1921; Howard and Anson, 1922; Haag, 2012).

(a)

(b)

FIGURE 2.10 (a) The mantle magazine of the pistolgrip (*Tritogonia verrucosa*). (b) The mantle magazine of winged mapleleaf (*Quadrula fragosa*).
Photos: (a) Matthew Patterson, USFWS. (b) Bernard Sietman, Minnesota Department of Natural Resources.

2.9   ATTACHMENT AND ENCAPSULATION

In glochidia, chemical and mechanical stimuli trigger the adductor mussel to contract and clamp the two valves onto the host tissue (Arey, 1921). Hooked glochidia appear to be more sensitive to tactile or mechanical stimulation; however, the shells may re-open if the mechanical stimulus is not followed up with some form of chemical stimulus (Lefevre and Curtis, 1910, 1912; Shadoan and Dimock, 2000). Hookless glochidia are more responsive to chemical stimuli (Lefevre and Curtis, 1910, 1912; Shadoan and Dimock, 2000).

Hooked glochidia tend to attach to the fins and other exterior tissues of the host (Barnhart *et al.*, 2008). External attachment exposes hooked glochidia to forces that could easily dislodge them from the host, so a strong grip is advantageous. A large diameter adductor muscle and depressed shell shape (decreased height to length ratio) are adapted for strong grip strength in the hooked larvae (Hoggarth and Gaunt, 1988). When the adductor muscle contracts, the hooks penetrate and sever the host tissue (Blystad, 1923; Arey, 1932; Fisher and Dimock, 2002).

Hookless glochidia attach primarily to the gill filaments (Figure 2.11) of the host (Lefevre and Curtis, 1912). Because fish gills are protected by the operculum, dislodgement of the glochidia is less likely and strong grip strength is not as important. Instead, hookless glochidia maximize the chances of contacting the host tissue by increasing the area swept when the valves close in response to stimulation (Hoggarth and Gaunt, 1988). A increased height to length ratio and a long adductor muscle (as opposed to the large diameter muscle) increase shell gape and sweep area (Hoggarth and Gaunt, 1988). When stimulated, the valves close, but do not sever the host tissues.

Tissue damage caused by valve closure initiates an immune response within the host, forming a capsule around the glochidium. Capsule is the preferred term (cyst is commonly used in the freshwater mussel literature) because the tissue surrounding the glochidium comes entirely from the host (Arey, 1921, 1932; Karna and Milleman, 1978). In fish, the process of healing damaged tissues occurs quickly to avoid pathogens and osmotic stressors present in the liquid

FIGURE 2.11 Glochidia of the plain pocketbook (*Lampsilis cardium*) attached to the gill filaments of a largemouth bass (*Micropterus salmoides*).
Photo: Molly Webb, USFWS. (A black-and-white version of this figure will appear in some formats. For the color version, please refer to the plate section.)

medium (Quilhac and Sire, 1999). In laboratory infestations, the bluegill (*Lepomis macrochirus*) completely encapsulated all larval paper pondshell (*Utterbackia imbecillis*) within 6 hours of attachment (Rogers-Lowery and Dimock, 2006). The speed of the process rules out the possibility that the host utilizes cell proliferation to encapsulate the glochidia (Arey, 1932). Instead, encapsulation involves migration of epithelial cells called keratocytes over the glochidium (Arey, 1921; Rogers-Lowery and Dimock, 2006). The presence of epithelial cells with a wave-like appearance supports the concept of encapsulation through cell migration (Jeong, 1989; Nezlin *et al.*, 1994).

If a glochidium attaches to an unsuitable host, the capsule still forms, but is enlarged and irregular in shape (Arey,1932; Fustish and Milleman, 1978; Waller and Mitchell, 1989; Rogers-Lowery and Dimock, 2006). The incompatible glochidium is attacked by the host's innate immune system and the larval tissue between the valves is destroyed (Arey, 1932). The shells are sloughed off the host before metamorphosis is complete (Rogers-Lowery and Dimock, 2006; Barnhart *et al.*, 2008). A suitable host can become unsuitable through the process of acquired immunity (Reuling, 1919; Arey, 1932; Rogers-Lowery and Dimock, 2006). Short-nosed gar (*Lepisosteus osseus*) and long-nosed gar (*Lepisosteus platostomus*) exposed to subsequent infections of the yellow sand shell (*Lampsilis teres*) produced juvenile mussels on the first and second infestation, but the glochidia were sloughed prior to metamorphosis on the third and all subsequent infestations (Reuling, 1919). Capsules formed on hosts with acquired immunity also appear enlarged and irregular in shape (Rueling, 1919; Arey, 1932; Rogers-Lowery and Dimock, 2006). Some evidence indicates that acquired immunity to one mussel species can lead to immunity against other mussel species (Rueling, 1919).

In contrast to glochidia, lasidia use hooks and ciliated lobes to attach to the host (Barnhart *et al.*, 2008). Once attached, the lasidia of some species go through the encapsulation process much like glochidia. Encapsulation of lasidia can be found in the genus *Anodontites* (Mycetopodidae). Other species are not encapsulated by the host, but instead metamorphose into a second larval stage that is attached to the host by a pair of haustoria (Wachtler *et al.*, 2001). The formation of haustoria can be found in the genus *Mutela* (Iridinidae). The haustorium, which provides both attachment and nutrition to the developing lasidium, extends away from the fish and more closely resembles a fungal hypha than anything molluskan (Fryer, 1961). The developing larvae are located at the distal end of the haustorium.

2.10 METAMORPHOSIS

If a glochidium attaches to a suitable host, metamorphosis to the juvenile stage occurs inside the capsule.

The larval adductor muscle disintegrates and is replaced by juvenile tissues and organs, including the anterior and posterior adductor muscles, foot, gill buds, and digestive system (Coker *et al.*, 1921; Waller and Mitchell, 1989; Fisher and Dimock, 2002). In hooked glochidia, the host tissue captured between the valves is broken down, providing nutrients to the developing glochidium (Blystad, 1923). In hookless glochidia, the host tissues are not broken down. Instead, host red blood cells are visible between the two valves and the intact host tissue acts like a placenta, providing nutrients for the developing larva (Blystad, 1923). Stable isotope analysis confirms that glochidia obtain nutrients from the host during encapsulation (Fritts *et al.*, 2013).

The duration of encapsulation and metamorphosis can be highly variable, both within species and among species. Water temperature also has a significant impact on the length of the encapsulation period, with decreases in water temperature leading to slower metamorphosis. Encapsulation can range from a few days all the way up to several months for species that overwinter on the host (Howard and Anson, 1922; Young and Williams, 1984b; Steingraber *et al.*, 2007).

In most species, the glochidium does not increase in size during the encapsulation period (Lefevre and Curtis, 1912). Species with miniature glochidia (<100 microns in length) and axe-head glochidia are the exception to this rule (Barnhart *et al.*, 2008). The glochidia of the western pearlshell (*Margaritifera falcata*), the freshwater pearl mussel (*Margaritifera margaritifera*), Spengler's freshwater mussel (*Margaritifera auricularia*), the mapleleaf (*Quadrula quadrula*), the winged mapleleaf (*Quadrula fragosa*), the fragile papershell (*Leptodea fragilis*), the scaleshell (*Leptodea leptodon*), the pink heelsplitter (*Potamilus alatas*), the fat pocketbook (*Potamilus capax*), the inflated heelsplitter (*Potamilus inflatus*), the pink papershell (*Potamilus ohiensis*), the fawnsfoot (*Truncilla donaciformis*), and the deertoe (*Truncilla truncata*) all have been shown to grow during encapsulation (Coker and Surber, 1911; Surber, 1912; Howard, 1914; Howard and Anson, 1922; Murphy, 1942; Young and Williams, 1984a; Cummings and Mayer, 1993; Roe *et al.*, 1997; Araujo and Ramos, 2001; Barnhart, 2001; Araujo *et al.*, 2002; Steingraber *et al.*, 2007).

For example, a sevenfold increase in length (60–420 microns) has been observed in the western pearlshell (*Margaritifera falcata*) during encapsulation (Murphy, 1942).

Lasidia that pass through the haustorial larval stage exhibit the largest size increase during host attachment. The lasidium of *Mutela bourguignati*, for example, grows from 200 microns up to 1500 microns before dropping off the fish host (Fryer, 1961). Metamorphosis occurs at the distal end of the haustorium in these species and in the case of *Mutela bourguignati*, the juvenile is already filter feeding prior to release (Wachtler *et al.*, 2001).

2.11 JUVENILE DROP-OFF

Once metamorphosis is complete, the newly developed juvenile mussel begins moving the valves and extending the foot to press against the walls of the capsule (Arey, 1932). The capsule walls also appear to be thinning during this same period (Arey, 1932). Once liberated from the capsule, the juvenile falls to the stream bottom, burrows into the substrate and begins to feed. Food particles collected through pedal feeding or deposit feeding have been shown to be an important food resource at this early life stage (Gatenby *et al.*, 1996). Growth rates can be highly variable at the juvenile stage, both among and within species. Water temperature, water chemistry, availability of food resources, and many other factors can impact growth rates, but there are some general rules of thumb to follow. Shell growth is much faster at the early life stages, slowing at the onset of sexual maturity, and thin-shelled species tend to grow faster than thick-shelled species.

To help prevent dislodgement from the substrate during high flow events, juvenile mussels of some species produce byssal threads that attach to rocks, sticks, or other heavy objects on the stream bottom (Figure 2.12). There is very little information available for survival in the wild at the early life stages; however, estimates of survival

FIGURE 2.12 Juvenile eastern lampmussel (*Lampsilis radiata*) bound together by a mass of byssal threads.
Photo: Amy Maynard, the Conservation Management Institute, Virginia Polytechnic Institute and State University.

from larvae to juvenile are in the order of 10^{-5} to 10^{-6} (Haag, 2002). Once mussels have passed through this bottleneck, survival seems to depend heavily on growth rate and lifespan, with fast-growing species having shorter lifespans and higher rates of mortality than slower-growing species.

3 Host Species Identification, Acquisition, and Captive Care

Tony Brady and Catherine M. Gatenby

At the time this book was published, the majority of freshwater mussel propagation facilities were producing juvenile mussels by attaching the larvae to a live host (*in vivo* metamorphosis), which is the focus of this chapter. Chapter 5 will discuss *in vitro* metamorphosis, a technique that can be used to produce juvenile mussels without a live host.

This chapter addresses how to test for and identify suitable hosts for the parasitic larval life stage of freshwater mussels; how to transport and care for hosts in captivity; biosecurity protocols to prevent disease outbreaks within a propagation facility, as well as prevent the spread of pathogens while collecting host species in the wild; and how to properly dispose of hosts after they have been used for mussel propagation. While the vast majority of freshwater mussels in the order Unionoida use freshwater fish as a host, the term "host species" was chosen for this chapter because at least one mussel species, the salamander mussel (*Simpsonaias ambigua*) uses a non-fish species, the mudpuppy (*Necturus maculosus*) as a host. The term host species, however, will often be interchanged with host fish throughout this book due to the dominance of fish as the primary host.

3.1 HOST SPECIES SELECTION

A few mussel species in the order Unionoida exhibit direct development to the juvenile stage inside the female marsupium (e.g. the green floater, *Lasmigona subviridis*). The remaining species require attachment to a suitable host in the wild to complete metamorphosis (see Chapter 2). Consequently, one of the first steps in freshwater mussel propagation is selecting a suitable host species.

Numerous studies have been published on suitable host fishes for a wide array of freshwater mussel species. Additionally, *Ellipsaria*, an online newsletter publication of the Freshwater Mollusk Conservation Society, reports results of ongoing research, including host fish studies. Due to the large and ever-growing number of publications on freshwater mussel–host species associations and the variation in methods used to identify the host species (natural infestation with no confirmed metamorphosis, natural infestation with confirmed metamorphosis, laboratory infestation with no confirmed metamorphosis, and laboratory infestation with confirmed metamorphosis; Watters *et al.*, 2009), this book will not provide a list of recommended host species for propagating any given species of freshwater mussel. Instead, this book will provide a few recommendations to help the reader select a host species based on a search of the existing literature.

Reports of mussel–host relationships reported in the literature that do not confirm larval metamorphosis to the juvenile stage should be viewed with caution (Watters *et al.*, 2009). Mussel larvae can attach to an unsuitable host species in both natural and laboratory infestations. An enlarged and irregular capsule is still formed around the larval mussel, but the tissues are destroyed and the shells fall off the host before metamorphosis is complete (Arey, 1932; Rogers-Lowery and Dimock, 2006; Barnhart *et al.*, 2008). Natural and laboratory infestations in the literature that do confirm metamorphosis can help identify potential hosts for a given mussel species; however, if no data on the number of newly metamorphosed juveniles per fish are reported, it is difficult to assess if the potential host is a primary host or marginal host for propagation.

A primary host will consistently yield large numbers of juvenile mussels per fish, whereas a marginal host, while still allowing for metamorphosis to the juvenile stage, typically produces smaller numbers (Haag and Warren, 1997; O'Brien and Williams, 2002). A primary host, however, could have a low tolerance for the stress of captivity compared to a marginal host. Sauger (*Sauger canadensis*), for example, produce 10 times the number of juvenile black sandshell (*Ligumia recta*) than any other host (Khym and Layzer, 2000), but typically have a low tolerance for captivity. In the absence of ideal holding conditions, sauger can experience high mortality before the glochidia complete metamorphosis. Infesting an intolerant host like sauger with a gill parasite will further compound any stress related to captivity. In this case, a marginal host with a higher tolerance for captivity may be a better choice. Largemouth bass (*Micropterus salmoides*), for example, are relatively tolerant of captivity. Indeed, hatchery-raised fish tend to have higher survival in captivity than fish collected from the wild. Lastly, if a primary host is difficult to collect from the wild in large numbers or is unavailable from a local hatchery, it may be necessary to use a marginal host that is easier to obtain.

When searching the literature, be aware that a host species identified using one mussel population may not be compatible with the same mussel species from a different geographical region. Riusech and Barnhart (2000), for example, showed that rainbow darters (*Etheostoma caeruleum*) that were sympatric with the Plea's mussel (*Venustaconcha pleasii*) produced more juveniles than rainbow darters from geographically distant populations. It may be necessary, therefore, to confirm the suitability of a host species identified in the literature with laboratory testing. Section 3.2 will cover host species testing.

3.2 HOST SPECIES IDENTIFICATION

If the host for a mussel species is unknown, two methods can be used to identify the primary host. One method is to collect fish from the wild where the mussel species is known to occur and search for natural infestations of mussel larvae. Naturally infested fish are returned to the laboratory and placed in aquaria until the glochidia complete metamorphosis. A couple of techniques are then available for identifying the newly metamorphosed juveniles. Allen *et al.* (2007) identified newly metamorphosed juveniles by measuring valve length, valve height, and hinge length with the aid of scanning electron micrographs. Juvenile measurements were compared to known characters of glochidia in the same drainage. Newly metamorphosed juveniles have also been identified using DNA barcoding technology, although Boyer *et al.* (2011) advocate using both DNA analyses and morphological characters for proper identification.

The more widely used method is host species testing in the laboratory. Host species testing begins by compiling a list of all the fish species that are sympatric with a particular mussel species. This initial list can be narrowed down by searching the literature for host relationships for mussel species in the same genus. Closely related species oftentimes use a similar suite of hosts. Basic information about the host attraction strategy also can help narrow down the list of potential hosts. If a mussel species uses a large mantle lure (tribe Lampsilini), the host is likely to be a large predator (e.g. black bass, walleye, etc.). If a mussel species uses a pelagic conglutinate (tribe Pleurobimini), the host is likely a cyprinid or other species that feeds on small particles in the water column. The most common families of freshwater fish that serve as hosts for North American freshwater mussels include Cyprinidae, Ictaluridae, Centrarchidae, Percidae, Cottidae, Sciaenidae, and Salmonidae (Fuller, 1974; Hoggarth, 1992).

Once the list has been narrowed down, specimens of each host species can be collected from the wild or purchased (see Section 3.3) and infested with glochidia (see Section 5.6). Each potential host species must be isolated in a separate tank to determine which species support metamorphosis to the juvenile stage and which species produce the largest number of juveniles per host. Static aquaria can be used to hold infested fish in isolation, but the daily upkeep required for maintaining water quality can be labor intensive. Missouri State University was one of the first facilities to start using zebrafish or *Xenopus* multi-tank recirculating systems (AHAB) for host species testing and juvenile propagation (Figure 3.1a). The multi-tank systems, originally designed for holding aquatic organisms in the laboratory for research, can be outfitted with biological filtration, ultraviolet filtration, temperature control, and oxygenation to maintain water quality.

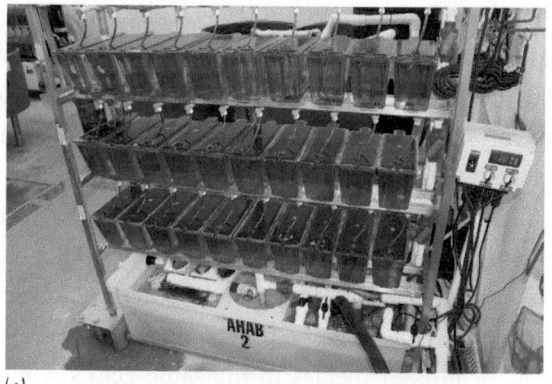

(a)

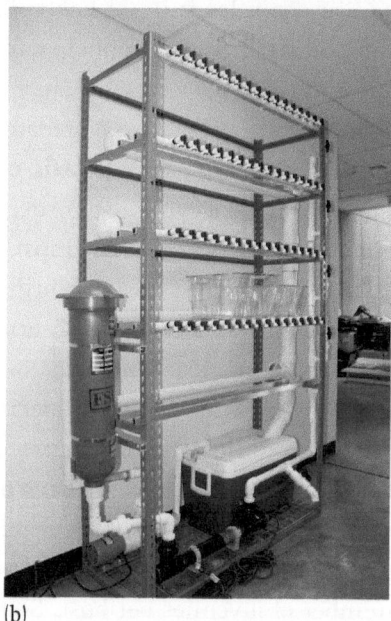

(b)

FIGURE 3.1 (a) A pre-fabricated multi-tank system (AHAB) used for mussel propagation and host species testing at the Virginia Fisheries and Aquatic Wildlife Center at Harrison Lake National Fish Hatchery (Charles City, Virginia). This system is outfitted with a series of 1 liter tanks with individual flow control for each tank, temperature control, and biological, mechanical, and ultraviolet filtration. (b) A homemade multi-tank recirculating system constructed in the laboratory at Missouri State University. The cooler serves as the sump and the cartridge filter and UV filters clean the water before it enters the individual tanks. (c) Ball valves provide individual flow control to each tank in the multi-tank system. (d) Individual tanks in the multi-tank system drain into a PVC trough that sends water back to the sump.
Photos: (a) Matthew Patterson, USFWS. (b, c, d) Chris Barnhart, Missouri State University.

(c)

(d)

FIGURE 3.1 (cont.)

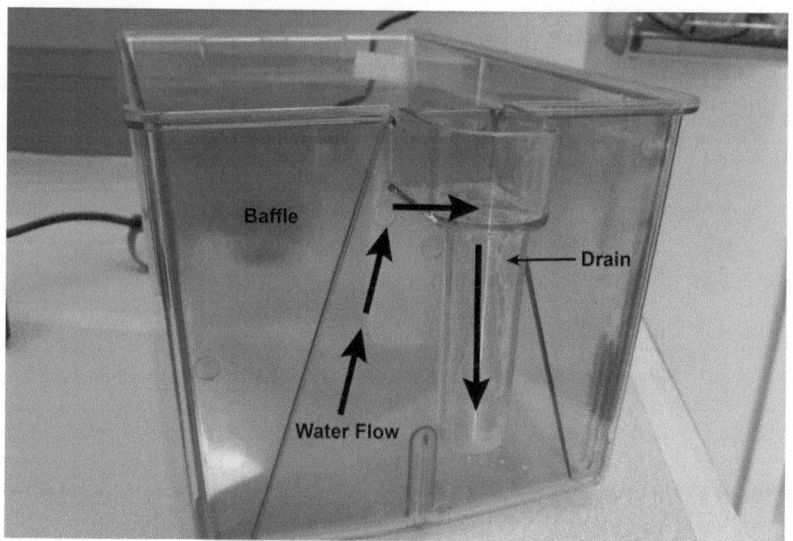

FIGURE 3.2 A self-cleaning tank from a multi-tank system. This view from the back of the tank shows the direction of the water flow. The water must exit under the baffle, creating water currents along the bottom of the tank that entrain juvenile mussels.
Photo: Matthew Patterson, USFWS. Graphics: Kristin Simanek, USFWS.

Temperature control (in either static or recirculating tanks) allows for manipulation of the rate of juvenile metamorphosis. Commercially available multi-tank systems range in price from \$10 000 to \$35 000; however, similar systems can be constructed by hatchery staff at a fraction of the cost (Figure 3.1b, c, d).

Each tank in a multi-tank system is designed to be self-cleaning. A rear baffle creates water currents along the bottom of the tank that remove fish waste (Figure 3.2). Newly metamorphosed juvenile mussels are very light and become entrained in the tank outflow along with the waste. Consequently, the outflow of each tank (or group of tanks if they are holding the same host-species–mussel pairing) must be outfitted with a fine-mesh filter to capture the juveniles before they are carried to the sump. Outflow filters currently in use in the United States include small plankton nets or filter socks

(Figure 3.3a, b), filter cups (Figure 3.3c, d, e), and nested sieves (Gatenby, 1994; Figure 3.3f, g). The filter socks and filter cups are simply placed over the outflow pipe to collect juveniles. The nested sieves use a series of filters that gradually decrease in mesh size. The course mesh filter on top (300 microns) prevents larger particles (e.g. fish feces) from contaminating the fine mesh filter on the bottom. The removal of the larger waste particles makes enumeration of the juvenile mussels under the microscope much easier and more accurate. The mesh on the plankton net or the last filter in the series of nested sieves must be smaller than the expected size of the juvenile mussels (125–250 microns, depending on the mussel species). The dimensions of some North American glochidia are described in Williams *et al.* (2008) and Watters *et al.* (2009).

All fish holding tanks and juvenile collection sieves should be monitored periodically throughout the encapsulation period for the presence of newly metamorphosed juveniles. How soon after the infestation process to begin monitoring will depend on the mussel species and water temperature. Some species complete metamorphosis in a few days, while species that overwinter on the host can take several months (Howard and Anson, 1922; Young and Williams, 1984b; Steingraber *et al.*, 2007). Warmer water temperatures increase the rate of metamorphosis and decrease the duration of encapsulation. If water temperatures are relatively cool (20 °C or less), then post-infestation monitoring can take place once a week for the first couple of weeks. Temperatures greater than 20 °C may require more frequent monitoring. After two weeks, monitor host tanks daily or every other day until the first juvenile mussels are collected.

Static aquaria are monitored by siphoning the bottom of the tank through nested sieves. Self-cleaning tanks are monitored by examining the contents of the nested sieves or the filter sock. Manual siphoning of the self-cleaning tanks can help ensure all juveniles are swept from the tank. If the self-cleaning tank set-up consistently removes all juveniles, however, manual siphoning may not be necessary. If the fine mesh filters on a self-cleaning tank begin to clog

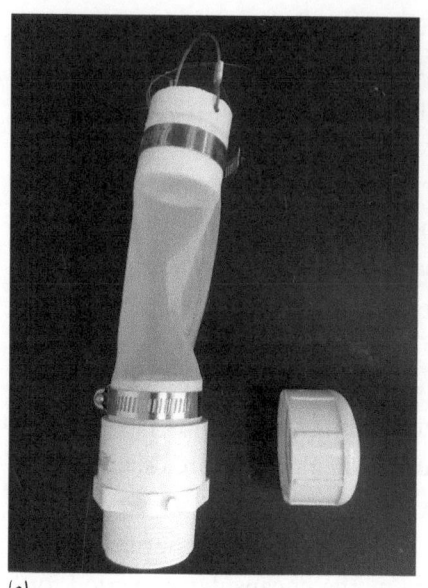
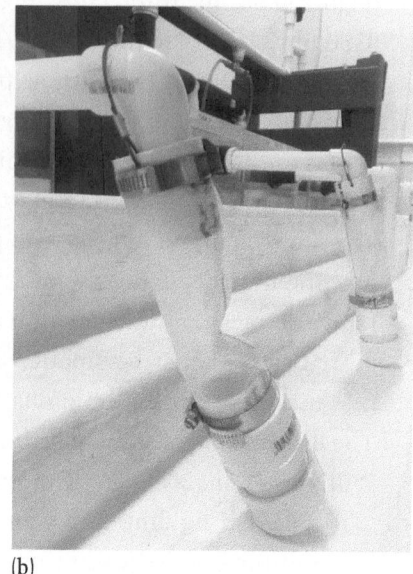

(a)   (b)

FIGURE 3.3 (a) A homemade filter sock made of PVC pipe and fittings, 125 micron mesh screening, and stainless steel hose clamps. The hose clamps need to be special marine grade stainless steel to prevent rust. (b) The homemade filter sock attached to the fish host tank outflow to capture juvenile mussels at the Orangeburg National Fish Hatchery (Orangeburg, South Carolina). (c) A filter cup used to collect juvenile mussels at Missouri State University. The cup is constructed of a short piece of PVC pipe with legs cut in the bottom to allow water to drain, fine mesh screening, and a PVC coupler. (d) The PVC pipe and PVC coupler are pressed together with the piece of fine mesh screen in between, forming a filter cup. (e) The filter cups are placed in the drain trough immediately below the outflow of the self-cleaning tanks to collect juveniles. (f) A pair of nested sieves attached to the fish host tank outflow for collecting juvenile mussels at the Virginia Fisheries and Aquatic Wildlife Center at Harrison Lake National Fish Hatchery (Charles City, Virginia). The 300 micron sieve serves as a pre-filter to collect fish feces or uneaten feed. The 150 micron sieve collects the juvenile mussels. (g) Each nested sieve is constructed using a PVC male adapter with 150 or 300 micron mesh screening glued to the bottom with PVC glue. The reduced diameter of the threaded portion of the male adapter allows the sieves to nest.

Photos: (a, b) Jonathan Wardell, USFWS. (c, d, e) Chris Barnhart, Missouri State University. (f) Rachel Mair, USFWS. (g) Matthew Patterson, USFWS.

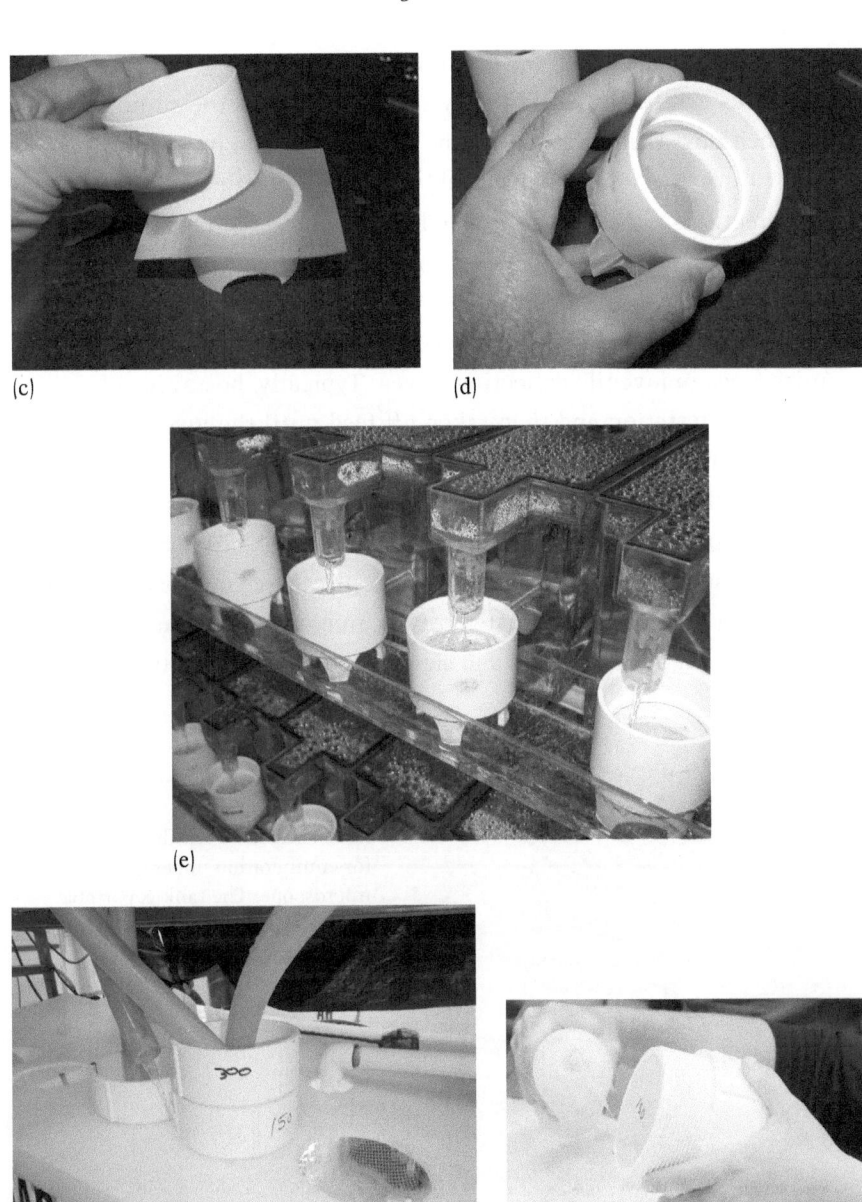

(c)

(d)

(e)

(f)

(g)

FIGURE 3.3 (cont.)

and overflow, the frequency of monitoring will have to be increased. Juveniles can be lost from the system, escape into the sump or get pumped into a different tank if the filter overflows. Juvenile escapement into an adjoining tank could lead to an unsuitable host being incorrectly identified as a suitable host or vice versa. Outflow filters should be cleaned and examined for juveniles frequently to prevent clogging. Taking the host species off feed will help remove waste materials from the system and prevent filter clogging, as well as minimize feces in juvenile collecting sieves. Typically, hosts are fed well prior to infestation and then taken off feed until the juveniles have stopped dropping off of the fish (Gatenby, 1994).

Wash all the material collected from the plankton nets or nested sieves into a Petri dish (or other container) using a squirt bottle or pressurized garden sprayer (Figure 3.4). Then place the Petri dish under a dissecting or cross-polarizing stereoscopic microscope to enumerate the juveniles. All outflow filters and sieves should be washed thoroughly

FIGURE 3.4 A pressurized garden sprayer used for transferring juveniles from a sieve or filter sock to a Petri dish for enumeration under the microscope. The tank is portable and can be pressurized by hand. Photo: Matthew Patterson, USFWS.

before moving to the next tank, as some mussels may not rinse out into the Petri dish. This is especially important during host testing. Remember to wash and sterilize equipment such as nets, sieves, and siphon hoses after use in each tank to prevent the spread of disease.

Even under ideal conditions, some of the hosts may die during the encapsulation period. To ensure a successful host test, be sure to infest multiple individuals of each host species. The ideal number of hosts will depend on the availability of the host and the size of the aquarium system. If only a few individuals of a given host can be collected for testing, keep them in separate tanks. If one individual becomes sick, the illness will not spread to other hosts. Even if hosts are plentiful, holding them in at least two different tanks will help prevent complete losses of fish if one tank is contaminated.

3.3   HOST SPECIES ACQUISITION

Once a primary host has been identified, purchasing fish from a commercial hatchery can be an easy way to acquire large numbers of fish for mussel propagation. Hatchery-reared fish are conditioned to handling and are typically more tolerant to the stress of captivity and larval infestation. Commercial hatcheries, however, may not have the appropriate size of fish or fish species available year-round. For example, smallmouth bass (*Micropterus dolomieu*) may be only fingerling size in the fall when the mussels are gravid. Smaller host species like minnows, darters, and sculpins are rarely available from a commercial hatchery. Coordination well in advance is important to ensure a hatchery will have the right species and size available when the mussels are gravid. Commercial hatcheries may not be able to deliver fish when needed or they may charge a fee for delivery. The mussel propagation program, therefore, may want to invest in a fish-hauling truck for delivering host fish.

Host fish also can be obtained from federal or state hatcheries in the United States, and many hold up-to-date fish health certifications, including Class A disease-free fish health certifications. Like commercial hatcheries, federal and state hatcheries can produce large

FIGURE 3.5 Collecting wild fish for freshwater mussel propagation using a backpack electrofisher, net, and seine.
Photo: Matthew Patterson, USFWS.

numbers of fish and provide sufficient numbers for a mussel propagation program. Regardless of where you plan to obtain host fish, it's important to have a back-up plan for when fish are not available.

If a primary host is not available from a hatchery, it will have to be collected from the wild (Figure 3.5). One advantage to using wild fish is the ability to collect them on short notice just prior to glochidia maturation. Good advance planning will help limit the need to collect host fish at the last minute, but sometimes it cannot be avoided. Last-minute host fish collection could be impacted by changing weather conditions or rising water levels, so it is important to have a back-up plan in place.

The primary drawback of using wild fish is the potential negative impact to wild fish populations. It is good practice to identify multiple locations for collecting a host species and rotate the collecting sites to minimize impacts to the host population.

In the United States, a scientific collecting permit is required when collecting fish from the wild. This chapter will not cover all techniques for collecting fish in the wild because numerous books and publications are available on this topic. It is important to note, however, that electrofishing may not be the best collection method for some commonly used host species. For example, special care must be taken when collecting small members of the family Cyprinidae (minnows and shiners). While some groups like the dace may be relatively tolerant of electroshocking, others groups like the shiners may not do well in the hatchery after shocking. Consider collecting less-tolerant species with seine halls to increase survival in captivity.

If fish are being held in a 5 gallon bucket during collection, water in the bucket should be exchanged frequently. Water exchanges will help maintain oxygen concentrations and limit large temperature fluctuations, especially during the warmer months. To minimize overcrowding in the 5 gallon bucket, periodically transfer fish to a larger holding system such as a large cooler or fish-hauling tank (Figure 3.6a, b), especially when air temperatures exceed 26 °C. Water in the larger holding system should also be exchanged on a regular basis to maintain temperature and oxygen concentrations. Because oxygen consumption increases as fish excitement levels increase during collection and handling, use portable aerators in the coolers or tanks and fish-hauling systems, especially if the holding time is greater than 1 hour.

If possible, avoid collecting host fishes in multiple drainages in the same day. If multiple drainages must be visited on a single trip, be sure to have dedicated equipment for each site (e.g. nets, waders, electroshocking gear, etc.) or decontaminate all equipment before proceeding to the next site.

3.4 HOST SPECIES TRANSPORT

The primary concern when a host species is being transported to the propagation facility is stress. Stress can cause mortality during transport or lead to disease outbreaks after the hosts arrive on station.

(a)

(b)

FIGURE 3.6 (a) A small cooler can be used to transport small host fish back to the propagation facility. (b) Transporting larger host fish or large numbers of host fish to the propagation facility may require a large fish-hauling truck.
Photos: Matthew Patterson, USFWS.

A stressed fish also may not survive the infestation process. The best way to minimize stress in the transport tank is to maintain cool water temperatures, maintain adequate dissolved oxygen, and keep the fish in darkness.

Maintaining dissolved oxygen concentrations above 5 ppm at all times is critical to minimizing stress during fish transport. Oxygen concentrations in the transport tank are affected primarily by water temperature and the weight of the fish. Transporting fish in cold or cool water has two primary benefits: (1) colder water holds more dissolved oxygen and (2) oxygen consumption decreases with decreasing water temperatures. Oxygen consumption in fish, for example, drops by 50% with every 10 °C drop in water temperature (Piper *et al.*, 1982). In case the primary oxygen delivery system fails or traffic problems increase transport time back to the facility, carry back-up equipment for maintaining dissolved oxygen.

Fish biomass during transport also can affect stress levels. Noga (1996) suggests using 1 liter of water for every 1 cm of fish. Berka (1986) recommends that the ratio of fish volume to transport water volume should not exceed 1:3. If supplemental oxygen is available during transport, fish biomass can be increased. Salt at a concentration of 1000 to 3000 ppm (3.8 to 11.4 g/gallon) can be added to the transport water to help minimize osmoregulatory stress and limit disease outbreaks (Francis-Floyd, 1995). Transporting fish in the dark also can help reduce stress.

Finally, measure the water temperature at the receiving facility and the collection site before heading out to the field. If water temperatures at the receiving facility can be heated or chilled to match the collection site, tempering may not be necessary. If water temperatures cannot be manipulated and the receiving water is warmer than the collection water, tempering may be needed. If the receiving water is colder, spring water ice or other non-chlorinated ice can be used to cool the water during transport.

Upon return to the propagation facility, all fish collection and transport equipment should be cleaned and disinfected. Contaminated equipment could transport fish diseases and invasive species to new areas, especially if the next collecting trip will occur in a different drainage or watershed.

3.5 BIOSECURITY AND QUARANTINE

Bringing wild or hatchery-reared fish onto an existing hatchery or other facility has the potential to put both long-standing and new fish production programs at risk. Comprehensive and clear guidance on biosecurity is critical. To avoid contamination of existing hatchery programs, consider the following: building the mussel propagation facility in a location that can be isolated from the main hatchery, creating separate hatchery entrances for hauling trucks and hatchery personnel for both the fish program and the mussel program, purchasing dedicated equipment (e.g. waders, nets, etc.) for the mussel propagation facility, labeling dedicated equipment to help prevent transfer to other areas of the hatchery (Figure 3.7a), building quarantine areas for any wild fish (and wild mussels) coming on station, creating and posting strict disinfectant protocols for all equipment and personnel, installing chemical footbaths (Figure 3.7b) and biosecurity signage at all building entrances (Figure 3.7c), and installing separate discharge pipes for mussel propagation infrastructure.

If a host is obtained from a state or federal hatchery that undergoes periodic fish health inspections (Figure 3.8), a health certificate may be available. If a host will be collected from the wild or obtained from a hatchery that does not undergo regular disease testing, contact a local fish health laboratory to request assistance. Disease testing typically requires a minimum of 60 individuals for each species tested. Test results can take 30 to 60 days to complete so be sure to plan for any disease testing well in advance. If the host is rare and 60 individuals are not available for testing, talk to a fish health professional to discuss possible alternatives to the standard minimum for health assessments.

Even if a health certificate is available, it is still good practice to assume that any host coming on station could be a potential vector of

(a)

(b)

FIGURE 3.7 (a) Labeling all quarantine equipment with colored tape can help prevent transfer of that equipment to clean areas of the propagation facility. (b) Disinfecting footbaths and disposable booties can be used to help prevent the spread of pathogens at the propagation facility. (c) Proper signage also helps maintain biosecurity at the propagation facility.

Photos: Matthew Patterson, USFWS. (A black-and-white version of part (a) of this figure will appear in some formats. For the color version, please refer to the plate section.)

(c)

FIGURE 3.7 (*cont.*)

disease. All incoming hosts should be isolated in a quarantine facility, quarantine room, or quarantine tank. The quarantine area should have dedicated nets, buckets, feeders, siphons, thermometers, and other supplies clearly labeled quarantine only. If multiple sets of equipment are not feasible, then all equipment coming in contact with a quarantine tank must be properly disinfected (e.g. Clorox, Virkon, etc.) prior to use in a "clean" tank. Hand-sanitizing stations and chemical foot-baths should be used when personnel travel between the quarantine area and any other part of the facility. As an additional precaution, dedicated boots and waders could also be used while working in the quarantine area.

Isolation in quarantine can help limit the spread of any incoming pathogens to "clean" areas of the facility. Fish can also be monitored

FIGURE 3.8 US Fish and Wildlife Service personnel completing a fish health inspection at the Southwestern Native Aquatic Resource and Recovery Center (Dexter, New Mexico).
Photo: Matthew Patterson, USFWS.

and treated during the quarantine period to ensure they are healthy enough for the infestation process. Infesting unhealthy fish could lead to mortality and the loss of an entire cohort of glochidia. Most zoos and aquaria quarantine new fish for a minimum of 30 days (Hadfield and Clayton, 2011); however, no set guidance for how long to quarantine host fish currently exists. Once a host clears the quarantine process and is declared healthy, it can be used for mussel propagation.

None of the water used to collect host species should be discharged outside the drainage in which they were collected without proper treatment. Similarly, water used to hold fish that are from outside the facility's drainage area also should be treated prior to discharge. Improper discharge of water could lead to the spread of pathogens and non-native species. Fish collected from different drainages should be isolated in different tank systems, in addition to going through a quarantine period.

3.6 MAINTAINING HOST SPECIES' HEALTH IN CAPTIVITY

Maintaining the health of the host species both during quarantine and post-infestation is critical to ensuring the larval mussels complete metamorphosis. A quality diet is essential for maintaining the health of potential hosts in captivity. Prior to infestation with glochidia, host species should be well fed using standard protocols for each species. Some hatchery-raised fish such as trout or bass may already be trained on commercially available pellet feeds; while others like sauger or walleye may require live feed. Many small fishes like darters, sculpins, and minnows take well to frozen blood worms, brine shrimp, mysis shrimp, flake food, and live black worms (Figure 3.9). Fish can be held off feed for at least 1 week prior to the start of juvenile mussel drop-off (Gatenby, 1994).

FIGURE 3.9 Brine shrimp culture facility at Genoa National Fish Hatchery (Genoa, Wisconsin).
Photo: Ryan Hagerty, USFWS.

Maintaining good water quality also is essential to minimizing stress on the host. The most important water quality variables to monitor are dissolved oxygen, temperature, suspended solids, pH, ammonia, nitrite, CO_2, alkalinity, hardness, and total suspended solids (Timmons and Ebeling, 2010). The most critical parameter to measure is dissolved oxygen. For warm water fishes, dissolved oxygen concentrations should be kept above 5 ppm. Be aware that as temperature and salinity increase, the solubility of oxygen decreases.

Because fish are cold-blooded animals, water temperature affects a whole suite of physiological processes, including growth, reproduction, and respiration and feeding efficiency. Water temperature also affects the concentration of dissolved oxygen in the water. Temperature tolerances vary among different fish species; thus, refer to the literature for information on temperature ranges for various host species.

Ammonia, nitrite, and nitrate are nitrogenous waste products excreted by live fish or released from dead organisms and uneaten feed. These toxic waste products must be removed from the water. Ammonia occurs in two forms, un-ionized (NH_3) and ionized (NH_4^+) and the sum of the two is called total ammonia-nitrogen (TAN). The relative concentration of un-ionized ammonia, the most toxic form of ammonia, depends on pH, salinity, and temperature. As temperature, pH, and salinity increase, un-ionized ammonia increases. For long-term exposure, un-ionized ammonia should be below 0.05 ppm and TAN should be below 1.0 ppm (Timmons and Ebeling, 2010). Nitrite levels as low as 0.5 ppm can cause stress in fish. The ideal nitrite concentration is between 0.0–0.2 ppm and concentrations above 1.6 ppm can be lethal. Chloride ions in the water reduce nitrite absorption so the common treatment for high nitrite concentrations is the addition of salt. Nitrate is the least toxic of the nitrogenous waste products and concentrations can typically be kept at safe levels with regular water changes.

Optimal pH levels for the growth and survival of most aquatic organisms is between 6.5 and 9.0. Exposure to pH levels outside this range can cause stress or even death, but more importantly these extreme pH levels can affect the toxicity of other chemicals in the water. An increase in pH will increase un-ionized ammonia, the most toxic form of ammonia. The nitrifying organisms in a biological filter also can be affected by pH extremes, thus affecting the conversion of toxic ammonia to nitrate.

Alkalinity measures the pH-buffering capacity of water and can be manipulated through the addition of sodium bicarbonate. The recommended alkalinity for most fishes depends on the pH and CO_2 concentration in the water, but in general concentrations between 75 and 200 ppm calcium carbonate are considered acceptable for fish culture (Wurts and Durborow, 1992). Hardness (oftentimes confused with alkalinity because it is also reported in ppm calcium carbonate) is primarily a measure of the calcium (Ca^{2+}) and magnesium (Mg^{2+}) ions in the water. Calcium and magnesium ions are important for bone and scale formation, blood clotting, and retention of sodium and potassium in the bloodstream. Depending on the fish species, recommendations for total hardness range from 20 to 300 ppm.

Suspended solids can damage fish gills, transmit pathogens, and impact a wide array of aquaculture system processes. While different fish species will have different tolerances for suspended solids, the recommended maximum level of total suspended solids (TSS) is 25 mg TSS/L and the recommendation for normal operation is 10 mg TSS/L (Timmons and Ebeling, 2010). All solids should be removed from the system as soon as possible.

Even with a high-quality diet and strict water quality monitoring, fish can still become stressed, contract disease and die. Keeping good records of holding conditions and mortality events for each species will provide key information for making improvements in the future. Unfortunately, limited research is available on the effects of fish therapeutants on encapsulated glochidia. Rach *et al.* (2006) investigated the effect of 60 minute exposures to 20 ppm chloramine-T, 2 ppm cutrine, 200 ppm formalin, 100 ppm hydrogen peroxide, and 20000 ppm sodium chloride on metamorphosis of plain pocketbook (*Lampsilis cardium*) glochidia encapsulated on largemouth bass

(*Micropterus salmoides*). None of the treatments tested significantly decreased metamorphosis relative to untreated controls. Treatment of channel catfish (*Ictalurus punctatus*) with oxytetracycline at a concentration of 25 ppm for 24 hours also showed no negative effects on encapsulated glochidia of the winged mapleleaf (*Quadrula fragosa*; Tony Brady, unpublished data). More research is needed on the effect of fish therapeutants on encapsulated glochidia. Test any new fish-health drug therapies on a small subsample of infested hosts to ensure that entire batches of juvenile mussels are not lost.

3.7 HOST SPECIES DISPOSAL

When all juvenile mussels have dropped off the host, it is important to follow all natural resource agency protocols regarding disposal of the host species. Never release fish or mussels outside of their native range. Some agencies may not allow fish to be released back to their native range if health certifications are not current and may ask that the fish be euthanized. Host fish can be held on station and re-infested with glochidia at a later date, but fish have been shown to develop acquired immunity after repeated infestations (see Chapter 2).

4 Brood Stock Collection, Transportation, and Captive Care

Julie L. Devers and Matthew A. Patterson

Freshwater mussel propagation at this point is not true captive propagation. Propagation facilities do not hold adult male and female mussels in captivity year-round as a source of future brood stock. Instead, spawning and fertilization is allowed to occur naturally in the wild and then gravid females (a female mussel that is brooding larvae in the gills) are collected from the wild and returned after the infestation process. Captive propagation may be an option in the future as technology progresses.

Before collecting gravid females from the wild, it is important to include information in the Propagation Plan regarding: (1) choosing a source population, (2) genetic variation of the source population, (3) method and timing of collection, (4) maintenance of health while in captivity, and (5) method and timing of release.

4.1 CHOOSING A SOURCE POPULATION

Choosing a source population is one of the most important decisions for any propagation program. If a mussel species is critically imperiled, careful planning will be needed to ensure recovery efforts do not negatively impact the viability of the source population. The source population should be monitored before and after gravid females are removed to ensure no negative impacts. If a mussel species is locally abundant, translocation of adults may be a better recovery strategy to minimize the effects of domestication or artificial selection associated with propagation (George *et al.*, 2009). Regardless of the size of a source population, genetic analyses should be carried out to ensure the source population has a similar genetic make-up to the target population, especially for augmentation projects. Genetic analysis also will help determine the number of gravid females to include in the propagation program to best represent the genetic diversity of the source population. See Jones *et al.* (2006a) for an excellent review of the genetic concerns associated with mussel propagation.

It is good practice, especially when working with imperiled species, to tag and keep detailed records on all gravid females used for propagation. Genetic sampling also can be used to "mark" progeny and track parentage. Genetic marking requires funding for the initial analysis, maintenance of a database, and subsequent analysis of individuals found in the wild. Two methods commonly used to collect genetic samples include tissue clips and buccal swabs (Figure 4.1). Tissue clips are typically collected from the mantle or foot tissue. Buccal swabs are collected by gently prying the mussel open and using a brush to swipe DNA from the foot tissue (Henley *et al.*, 2006). Henley *et al.* (2006) also recommend brushing the tissue at least six times to obtain an adequate sample. Tissue samples and buccal swabs should be maintained in 100% non-denatured alcohol. Tissue clips more reliably yield DNA and in smaller animals can be less invasive than abrading internal body parts of the mussel (Meredith Bartron, USFWS, personal communication).

4.2   COLLECTING GRAVID FEMALES

Before collecting gravid females from the source population, it is important to obtain the appropriate collecting permits. In the United States, species designated as threatened or endangered under the Endangered Species Act of 1973 require both a federal endangered species collecting permit and a state collecting permit. Permits typically require reports at the end of the year so it is important to keep detailed records of all collections.

As with fish collection, it is important to minimize stress on gravid female mussels during collection. Minimize air exposure as much as possible by keeping mussels in the water. If gravid mussels will be held in buckets on shore or in the boat for tagging and processing, the water should be shaded and replenished on a regular basis to

FIGURE 4.1 Collecting a buccal swab from a sheepnose (*Plethobasus cyphyus*) for genetic analysis.
Photo: Bryan Simmons, USFWS. (A black-and-white version of this figure will appear in some formats. For the color version, please refer to the plate section.)

maintain temperature and oxygen concentrations (especially during the warmer months). Water can be replenished periodically by manually carrying buckets of water from the river or by using a submersible pump to continuously pump water from the river (Figure 4.2a, b). Portable aerators also can be used to maintain oxygen levels, if available. Ideally, gravid females should be held in a shaded area of the river to keep them cool and well oxygenated prior to transport. Mussels

(a)

FIGURE 4.2 (a) Freshwater mussels being held in a large holding tank on shore during the measuring and tagging process. Water is continuously being pumped from the river using a small submersible pump to help maintain mussel condition. (b) The small submersible pump used to pump river water to the mussel holding tank.
Photos: Matthew Patterson, USFWS.

(b)

FIGURE 4.2 *(cont.)*

can be held directly in the river using laundry baskets (Figure 4.3a) or mesh bags (Figure 4.3b). The density of mussels in the mesh bags should be kept to a minimum, allowing the mussels to lay flat on the river bottom in a single layer. If multiple species are being collected, smaller, thin-shelled species should be kept in separate bags to prevent overcrowding and crushing by the larger, thick-shelled species.

Collecting gravid females at the right time of year is very important. Collection should occur when the brooded larvae are mature, so knowledge of the brooding cycle of each mussel species is critical. In general, the larvae of short-term brooders (tachytictic) are mature for a very short period of time in the summer and the larvae of long-term brooders (bradytictic) mature in the spring (Williams *et al.*, 2008; Watters *et al.*, 2009). Some species exhibit visual cues that indicate mature larval brooding, including mantle lures, inflated mantle magazines, and changes to the color and the degree of inflatedness of the marsupia. Search the primary literature and contact local mussel experts to better understand the best time to collect gravid females of the target species.

(a)

(b)

FIGURE 4.3 (a) Freshwater mussels being held in laundry baskets in the river. The holes in the basket allow fresh river water to pass across the mussels. The fresh river water maintains water temperatures and delivers food and oxygen to the mussels. (b) Freshwater mussels being held in collecting bags in the river. The mussels are laid flat on the bottom of the river in a single layer to maintain condition. Photos: Matthew Patterson, USFWS. (A black-and-white version of part (b) of this figure will appear in some formats. For the color version, please refer to the plate section.)

In addition to the short window for collecting mature larvae, short-term brooders also may release immature larvae (which are unusable for propagation) in response to handling stress or changes in water temperature. Short-term brooders may even release larvae in collection bags or buckets immediately after collection or during transport back to the propagation facility. Because of this tendency toward premature release of larvae, it is critical to separate different species of short-term brooders during collection and transport. It also may be necessary to keep individual mussels of the same species separate if the Propagation Plan calls for parentage-based genetic tagging. Individual gravid females can be physically separated in individual coolers or individual ziplock bags within the same cooler. If gravid females are not kept separate, it will be impossible to separate the glochidia once they have been released.

For those species that tend to release larvae in response to handling, it is important to have the suitable host species ready and waiting at the facility prior to collecting brood stock. Glochidia viability declines rapidly after release from the female, especially in warmer water temperatures (Jansen *et al.*, 2001). Larvae of *Margaritifera laevis*, for example, survived between 4 and 11 days at 10 °C, but for only 24 hours at 15 °C and 20 °C (Akiyama and Iwakuma, 2007). A short larval lifespan outside of the adult female leaves little or no time to acquire the host species. Long-term brooders are less likely to release larvae in response to handling or temperature changes, and can typically be held in captivity for a longer period of time. Long-term brooders spawn in the fall and brood the larvae over winter so gravid females can be collected in the fall. Research with the oyster mussel (*Epioblasma capsaeformis*), however, showed increased growth and survival of juveniles when propagated in the spring of the year, when glochidia were mature and naturally released (Jones *et al.*, 2005).

Methods used to collect gravid female mussels vary with stream size. In small streams, view buckets, snorkeling, clam rakes, and even feeling around in the substrate can be used to collect mussels (Figure 4.4). SCUBA gear is typically required in areas where the water

(a)

(b)

FIGURE 4.4 (a) Searching for gravid females with a view bucket. (b) Snorkeling allows the collector to get their face as close to the substrate as possible, increasing the chances of finding mussels. (c) Collecting gravid females using a clam rake reinforced with 0.5 inch mesh hardware cloth to catch smaller mussels.

Photos: (a, b) Matthew Patterson, USFWS. (c) Jonathan Wardell, USFWS.

(c)

FIGURE 4.4 (cont.)

is greater than 1 to 1.5 meters in depth (Figure 4.5). View buckets increase visibility by removing glare on the water surface and can be constructed by attaching a piece of plexiglass to the bottom of a five gallon bucket or purchased from a commercial vendor. Snorkeling gear typically includes a wetsuit, mask and snorkel, and wading boots, as well as weight belts or ankle weights in areas of high flow. Snorkeling gets the collector's eyes closer to the river bottom than a view bucket, increasing the probability of finding mussels. SCUBA diving also gets the collector's eyes closer to the river bottom, but diving can be a hazardous activity. A clam rake works well in silty or sandy substrates but would have limited application in cobble/rocky substrates (Morgan Wolf, USFWS, personal communication).

Gravidity can be checked by gently prying open the shell to determine if the gill tissues are inflated. The degree of gill inflation varies, depending on the duration of the brooding period. In the long-term brooders, the change in gill thickness during brooding can be

FIGURE 4.5 SCUBA is typically used to collect gravid females in deeper water.
Photo: Ryan Hagerty, USFWS.

quite dramatic (Figure 4.6). Gill inflation in short-term brooders is typically much less dramatic and can be more difficult to detect. The shell can be pried open using a variety of instruments. Fingernails may be sufficient to pry open a few smaller species, but the larger, heavier-shelled species will require an assistive device like a pair of reverse pliers. The ideal instrument will depend on the mussel species, the size of the mussel, and the relative thickness of the shell (Figure 4.7). The shells of some thin-shelled species are easily cracked if all of the prying pressure is exerted on one small area of the shell. In this case, the plier blades should be wide enough to distribute the pressure more evenly across a larger area of the shell. The ideal place to insert the pliers or other device is along the anterior-ventral margin of the shell where the foot is extended (Figure 4.8).

The plier blades should be shaved down to a thin plate for easy insertion between the two valves. It also can be helpful to have a headlamp or flashlight to illuminate the gill filaments. Please open

Non-Gravid Gill

(a)

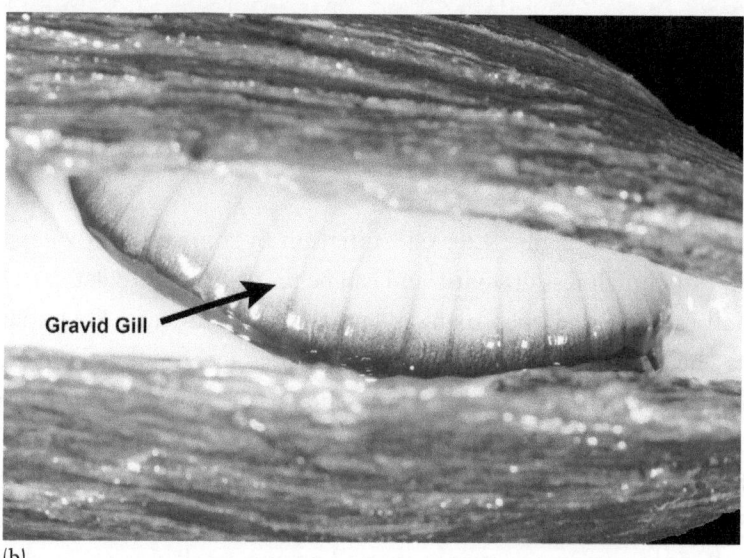

Gravid Gill

(b)

FIGURE 4.6 (a) The Eastern lampmussel (*Lampsilis radiata*) gently pried open to reveal the gills and check for gravidity. The flattened gills in this female indicate it is not gravid. (b) The inflated gills of a female plain pocketbook (*Lampsilis cardium*), indicating this female mussel is gravid and brooding glochidia. (c) A freshly dissected fatmucket (*Lampsilis siliquoidea*) showing the inflated outer gill and the flattened inner gill of a long-term brooder.

Photos: (a) Matthew Patterson, USFWS. Graphics: Kristin Simanek, USFWS. (b) Ryan Hagerty, USFWS. Graphics: Kristin Simanek, USFWS. (c) Matthew Patterson, USFWS. Graphics: Kristin Simanek, USFWS. (A black-and-white version of parts (a) and (b) of this figure will appear in some formats. For the color version, please refer to the plate section.)

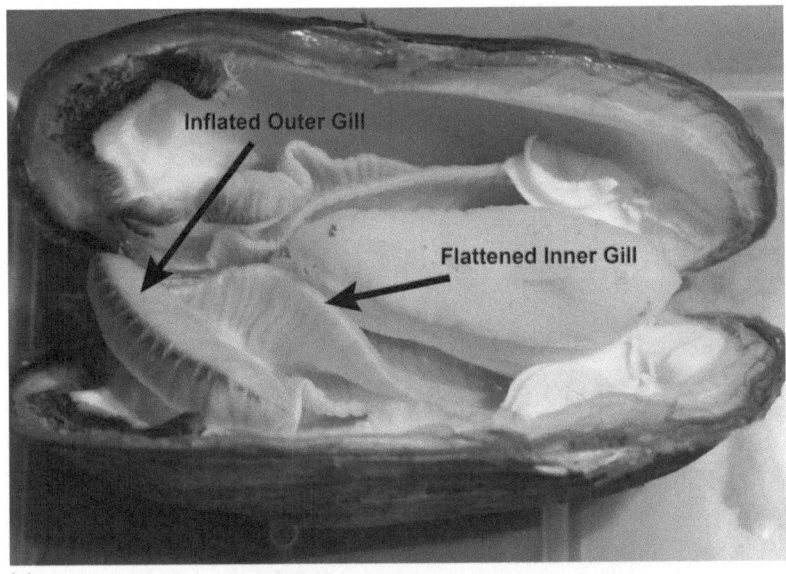

(c)

FIGURE 4.6 (cont.)

the shell slowly and be very careful not to use excessive force. It is easy to tear the adductor muscles, making it difficult for them to close the shell and possibly opening the mussel up to infection. If the adductor muscles are torn, the mussel will likely die.

For mussel species that tend to release undeveloped eggs or immature larvae in response to handling, larval maturity can be checked in the field. Maturity can be determined by removing five to ten larvae from the gills with a glass pipette and viewing them under a portable dissecting microscope. Immature glochidia have an incompletely formed shell and the soft tissue between the valves occupies greater than approximately one-third the total volume between the valves (Mark Hove, personal communication). Glochidia at this stage of development have so much soft tissue between the valves that little or no room is available for attachment to the host. Mature glochidia have fully formed valves and the volume of soft tissue between the valves is less than approximately one-third the total volume between the valves. Females bearing mature glochidia can be transported

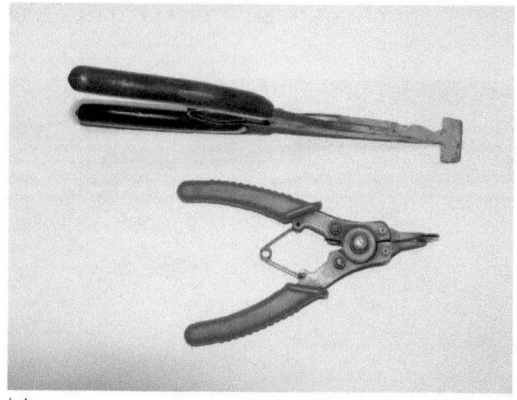

(a)

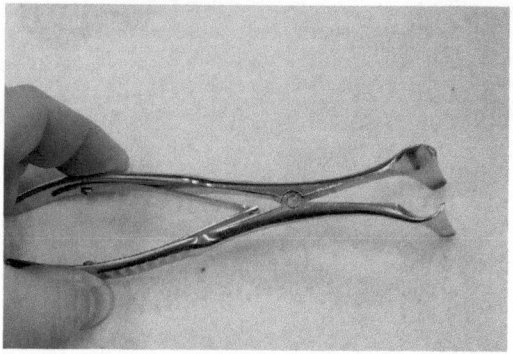

(b)

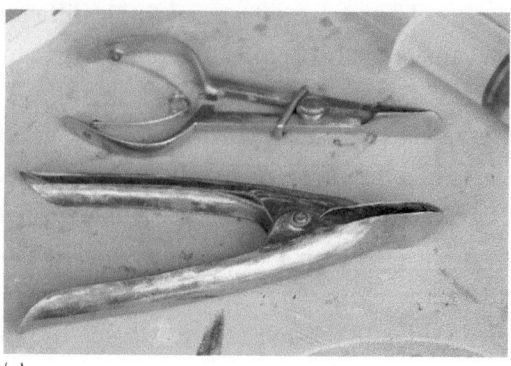

(c)

FIGURE 4.7 (a) Retaining ring reverse pliers modified for checking gravidity in freshwater mussels. The blades can be modified to make them easier to insert between the two valves. The blades of the top pair have been widened to distribute the force on the shell margin and minimize shell cracking. (b) A nasal speculum modified for checking gravidity in freshwater mussels. Squeezing the handles spreads the blades apart and opens the shell. (c) Reverse pliers used in the Chinese pearl industry for opening shells.
Photos: (a) Tony Brady, USFWS. (b) Matthew Patterson, USFWS.
(c) Ryan Hagerty, USFWS.

FIGURE 4.8 A nasal speculum being inserted along the anterior-ventral margin of the Eastern lampmussel (*Lampsilis radiata*). The blades of the speculum have been shaved down to make insertion between the shells easier.
Photo: Matthew Patterson, USFWS.

to the propagation facility while those bearing immature glochidia should be returned to the river, perhaps to be collected at a later date.

4.3   TRANSPORTING GRAVID FEMALES

Once a sufficient number of gravid females have been collected, it is time to transport them back to the propagation facility. The effects of transport on stress levels in freshwater mussels are relatively unknown. Marine bivalves under stress or in poor condition may still become gravid, but incubate fewer larvae (Walne, 1964). The eggs of stressed female bivalves may have decreased yolk reserves, resulting in reduced larval survival and vigor (Helm *et al.*, 1973; Bayne *et al.*, 1978). Taking steps to limit stress on gravid females during transport can improve survival in captivity and increase growth and survival of the progeny.

While the effects of emersion differ among species (Dietz, 1979; Chen *et al.*, 2001), air exposure during transport should be kept to a minimum. Air exposure has been shown to cause metabolic acidosis in freshwater mussels (Byrne and McMahon, 1991) and during the summer months lead to decreased survival in the threeridge (*Amblema plicata*) and threehorn wartyback (*Obliquaria reflexa*) compared to the fall (Waller *et al.*, 1995). Possible explanations for the decreased survival included seasonal temperatures, seasonal moisture levels, shell thickness, metabolic rates, and reproductive status. For example, high ambient air temperature and low relative humidity during air exposure both have been shown to decrease survival (Byrne and McMahon, 1991; Heath and Chen, 1996). Transporting freshwater mussels in air (in coolers wrapped in wet burlap) was found to be more stressful than transportation in aerated water (Chen *et al.*, 2001). Air transport, however, may be preferable to transport in water with low dissolved oxygen. The pondmussel (*Ligumia subrostrata*) survived 40 days in moist air but only 5–7 days in anoxic water (Dietz, 1979). Since most gravid females are collected from the wild in the warmer months of the year (spring and summer), care should be taken to avoid air exposure and mussels should be transported in water with vigorous aeration. Chlorine-free ice (or chlorinated ice in double ziplock bags) also can be added to the transport container to minimize temperature increases during transport.

4.4   BIOSECURITY

To limit the spread of aquatic invasive species, all gear that comes in contact with the water during collection should be cleaned and disinfected before moving to another collection location. Disinfection protocols can include bleach, salt, drying, or application of Virkon or a similar disinfectant (see Cope *et al.*, 2003 for chemical application rates and exposure times). If bleach, salt, or Virkon are used, all equipment should be thoroughly rinsed to ensure longevity of the equipment.

If gravid females will be collected from zebra-mussel-infested waters, special precautions must be taken to prevent their spread to the propagation facility or un-infested waters. All large, adult zebra mussels attached to the outside of the shell should be removed by hand. Juvenile zebra mussels and byssal threads should be removed with brushes or scrubbing pads. Tiny juvenile zebra mussels and veligers can be hidden in the crevices of old, damaged shells (broken hinge ligaments and cracks in the ventral margin), so special attention should be given to these individuals. Once scrubbed, all mussels should be transferred to clean water (zebra mussel/veliger-free) to prevent reinfestation by zebra-mussel veligers present in the river (Gatenby *et al.*, 2000; Cope *et al.*, 2003). Clean water should be replenished on a regular basis to maintain water temperature and oxygen concentration.

After transport to the propagation facility, all mussels collected from zebra-mussel-infested waters should be quarantined for a minimum of 30 days (Gatenby *et al.*, 2000). During the quarantine period, mussels should be supplied with a constant food supply to maintain good physiological condition. A constant supply of food also will enable small zebra mussels that went undetected during the initial scrubbing procedure to grow to sufficient size for detection. Quarantined mussels should be held in recirculating aquaculture systems to prevent the release of zebra mussel juveniles or veligers in flow-through effluents. Water removed from the quarantine tanks during regular water exchanges should be sterilized with 25 ppm chlorine (1.25 mL household bleach at 5.25% sodium hypochlorite per liter of water) for one hour prior to release into the environment or sewer system (Gatenby *et al.*, 2000; Cope *et al.*, 2003). After 30 days, each mussel should be inspected with a 4× magnifying glass and direct light. If zebra mussels are detected, all mussels should be scrubbed, placed in clean water (zebra-mussel-free) and quarantined for an additional 30 days. When the mussels are certified zebra-mussel-free, they can be removed from quarantine.

All quarantine water, tanks and pumps, and all collection and transport equipment (coolers, buckets, aerators, gear, etc.) should be properly sterilized before being released down the drain or used in non-zebra-mussel waters. Note that wastewater treatment systems are not designed to sterilize water so you should not expect them to kill all microscopic organisms. Water and equipment can be sterilized with a solution of 25 ppm chlorine for one hour. All chlorine residues should be neutralized with sodium thiosulfate. The amount of sodium thiosulfate needed and the exposure time needed to neutralize chlorine will vary with pH and water temperature. Chlorine test strips can be used to verify all chlorine residues have been neutralized.

4.5 MAINTAINING GRAVID FEMALES

Adult brood stock can be held in static, recirculating, or flow-through systems. The more closely a system resembles the natural environment, the better the chances of maintaining the condition of gravid females in captivity over the long term. In many cases, gravid females can be collected and returned to the river after a short stay in the hatchery. A high water event, however, may prevent immediate release back to the wild so it is wise to prepare for the possibility of holding gravid females for longer periods. If the propagation facility is located near a river that supports a native mussel population, a flow-through system that draws water from the river may be the best option. Wild river water will provide a wide variety of food resources and cut down on operational costs associated with producing algae on station. Flow-through systems also can support higher stocking densities because waste products are constantly being flushed from the system. Water tests should be run to ensure adequate water quality, temperature profiles, and food availability for the maintenance of condition. On the downside, flow-through systems can be vulnerable to natural and human-induced changes occurring in the river, including periods of low dissolved oxygen, chemical spills, drought conditions, etc.

If the propagation facility is not located near a river that supports a native freshwater mussel population, some form of static or recirculating system will be needed. The Virginia Fisheries and Aquatic Wildlife Center at Harrison Lake National Fish Hatchery

FIGURE 4.9 Refrigerated recirculating system for holding freshwater mussel brood stock at the Virginia Fisheries and Aquatic Wildlife Center at Harrison Lake National Fish Hatchery (Charles City, VA). Photo: Matthew Patterson, USFWS.

and Missouri State University hold gravid females in a refrigerated recirculating system (Figure 4.9). Water temperatures are maintained at 6 °C to minimize stress and prevent premature release of glochidia (Brian Watson, personal communication). Recirculating systems are not vulnerable to changes in the river mentioned above, but they are sensitive to the build-up of waste products and metabolites. Helm and Bourne (2004) recommend that the total live weight (including the shell) of marine bivalves held in recirculating systems not exceed

3 g/L. The authors also recommend a 100% water change twice per week to prevent the accumulation of metabolites. Research at White Sulphur Springs National Fish Hatchery showed that the mucket (*Actinonaias ligamentina*) can be held at densities higher than 3 g/L with no loss of condition (unpublished data in Morrison *et al.* 2013); however, these mussels were not being held for brood stock. The maximum of 3 g/L is likely a better guideline to follow for maintaining the condition of gravid females in recirculating systems.

If gravid females must be held at the facility for extended periods, they should be held in substrates suitable for burrowing. When burrowing in the substrate, freshwater mussels relax the adductor muscles and allow the expanding hinge ligament to open the valves (McMahon, 1991). Thus, providing substrate for burrowing will minimize energy expenditures associated with muscle contraction. Burrowing marine bivalve species also have been shown to feed more efficiently when provided with suitable substrate at the hatchery (Helm and Bourne, 2004). Optimal substrate depth will depend on the size of the mussel species, but should be deep enough to allow the mussels to burrow completely. Do not use substrate for holding short-term brooders. Any glochidia that are released can be extremely difficult to retrieve if substrate is present. Short-term brooders should instead be held in a dark-colored bin or trough inside the holding system that will make the detection and collection of glochidia easier.

Gravid females held for extended periods also require a constant supply of food. Cultured algae are the primary food supply for mussels held in recirculating systems. Algae can be grown on station or purchased commercially. Freshwater mussels are very efficient filter feeders so all the food resources can be removed very quickly during batch feeding (manual addition of food at the beginning and end of each day). With no food for extended periods, mussels may stop producing digestive enzymes. Significant energy expenditures will then be needed to create new enzymes when food again becomes available. This energy expenditure can lead to a loss of condition over time.

An automated feeding system that delivers feed on a continuous basis will be needed to maintain a constant food ration over a 24-hour period. Daily water samples should be collected from the holding system and analyzed with a Coulter Counter (or other device to measure algal cell densities) to ensure proper operation of the automated feeding system. Ideally, food rations should be based on organic matter concentrations from the site where the gravid females were collected. Feeding experiments with the mucket (*Actinonaias ligamentina*) collected from the Allegheny River showed that feeding a constant ration of 2 mg dry weight per liter maintained condition index (a standard method of evaluating condition in marine bivalves) for up to 8 months (unpublished data in Morrison *et al.*, 2013). Gatenby *et al.* (2013) recommended feeding 2.8 mg dry weight of algae (4.2×10^8 cells of the green alga, *Neochloris oleoabundans*) on a daily basis to optimize ingestion rate for adult rainbow mussels (*Villosa iris*).The marine bivalve industry conditions their brood stock at a rate of 2–4% of the mean dry tissue weight per day. Feed rates above 6% lead to brood stock growth at the expense of reproductive development (Helm and Bourne, 2004). The downside of providing a constant supply of food is the build-up of waste products and excess feed, especially in the substrate crevices. Recirculating systems will require frequent cleaning. Do not clean recirculating systems in the winter when energy reserves and metabolic rates are low. Winter handling can lead to significant mortality the following spring and summer.

Water temperature and flow rate in captivity should mimic river conditions where the gravid females were collected as much as possible. Mussel species from different river drainages should be kept in separate holding systems to prevent the transfer of pathogens and parasites. Similarly, if gravid females were collected from outside the hatchery watershed and water is being discharged to the local environment, all effluents should be treated prior to discharge. Effluents should be treated with 200 ppm free chlorine or other effective sterilizing agent for a minimum of 24 hours prior to discharge.

4.6 RELEASING GRAVID FEMALES

After larval extraction and host infestation (Chapter 5), all gravid females should be transported back to the river using the same guidelines discussed above to minimize stress. Prior to release, all gravid females should be tagged to track survival and genetic parentage. Tagging methods will be discussed in Chapter 7.

5 Larval Metamorphosis and Juvenile Mussel Collection

Nathan L. Eckert

Metamorphosis from the larval stage to the juvenile stage is one of the most important steps in freshwater mussel propagation. Metamorphosis in the laboratory can be completed either *in vivo* through infestation of a suitable host or *in vitro* in a suitable culture medium. This chapter will describe all aspects of the metamorphosis process from the harvesting of larvae through the collection of juvenile mussels.

5.1 HARVESTING LARVAE: LONG-TERM BROODERS

Many of the freshwater mussel species currently under propagation are long-term brooders (bradytictic). Propagation facilities have focused more on long-term brooders because they tend to be easier to propagate. Gravid females are easier to collect from the wild because of the extended brooding period. Long-term brooders are less likely to release immature larvae in response to handling and stress. The hosts for long-term brooders are typically more readily available. Finally, long-term brooders tend to brood the larvae loose in the marsupium, making harvest easier. The most common methods used for harvesting larvae from long-term brooders include the syringe method, the siphon method, the warming method, and chemical inducement. Table 5.1 lists the primary and secondary harvest methods used for a variety of freshwater mussel genera in North America.

5.1.1 *Syringe Method*

The most commonly used method for harvesting larvae from long-term brooders is the syringe method (Zale and Neves, 1982). To harvest larvae using the syringe method, first fill several syringes with

Table 5.1. *Larval extraction methods for a variety of freshwater mussel genera in North America*

Genus	Primary Method	Secondary Method
Actinonaias	Syringe	No secondary method needed
Alasmidonta		
Arcidens		
Ellipsaria		
Glebula		
Hemistena		
Lampsilis		
Lemiox		
Leptodea		
Ligumia		
Medionidus		
Obovaria		
Pegias		
Plectomerus		
Potamilus		
Pyganodon		
Simpsonaias		
Toxolasma		
Truncilla		
Uniomerus		
Utterbackia		
Venustaconcha		
Villosa		
Amblema	Isolation	Syringe
Anodontoides		
Elliptio		
Hamiota		
Lasmigona		
Obliquaria		
Megalonaias		
Epioblasma	Syringe	Siphon
Cyprogenia	Warming/Isolation	Chemical inducement
Dromus		
Ptychobranchus		

Table 5.1. *(cont.)*

Genus	Primary Method	Secondary Method
Anodonta	Warming/Isolation	Natural infestation
Cumberlandia		
Cyclonaias		
Fusconaia		
Margaritifera		
Plethobasus		
Pleurobema		
Pleuronaia		
Strophitus		
Quadrula	Suction	Isolation

clean water of the appropriate temperature (Figure 5.1). The water in the syringe should be free of fish mucus and the temperature should match the temperature of the system used to hold the gravid female. Next, fit each syringe with a hypodermic needle (Gauge 15 to 22). The ideal syringe volume, needle gauge, and needle length will depend on personal preference and comfort. Larger syringes will reduce the number of refills needed during the harvest procedure, but they can be more difficult to manage. It may be difficult to hold the syringe in one hand and depress the plunger while holding the gravid mussel in the other hand. Small syringes are easier to manage, but may require multiple refills to complete the harvest procedure. An auto-refilling syringe also can be used to avoid multiple refills of the syringe.

To open a gravid female mussel, insert reverse pliers between the two valves midway along the ventral margin of the shell, gently pry the valves apart, and insert a wedge to prevent the mussel from closing (Figure 5.2a). To avoid tearing the adductor muscles, do not spread the valves more than 0.25–0.50 inch (6–12 mm). Shine a head-lamp or flashlight between the two valves to obtain a clear view of the gills. Hold the mussel over a clean container with the ventral margin facing up and the anterior end facing away from you to ensure that all

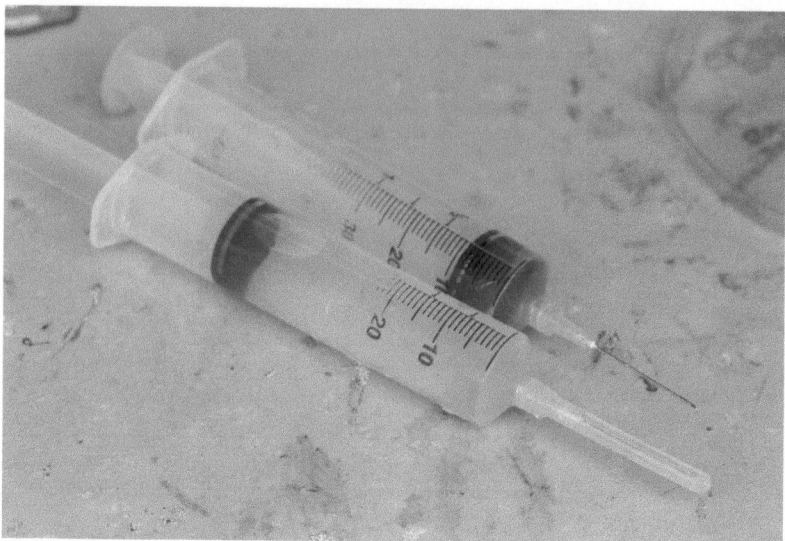

FIGURE 5.1 50 milliliter syringes filled with clean water for harvesting glochidia using the syringe method.
Photo: Ryan Hagerty, USFWS.

of the water and larvae drip directly into the container (Figure 5.2b). Larvae that land on the table or other surface likely will not attach to the host. Insert the hypodermic needle into the posterior end of the inflated gill, perpendicular to the water tubes (Figure 5.2c). Push the needle all the way to the anterior-most portion of the gill and slowly depress the syringe plunger to inject water into the gill (Figure 5.2d). While water is flowing into the gills, slowly pull the needle back toward the posterior portion of the gill in a circular motion to harvest the larvae from all the water tubes. Use a squeeze bottle to flush the mantle cavity with clean water to remove any larvae loose inside the shell (Figure 5.2e). When all of the larvae have been harvested, the gill should lay flat (Figure 5.2f). A chemical spatula or Huber probe can be used to inspect the gill for any water tubes that may have been missed. When all of the gills have been emptied, remove the wedge and return the mussel to the holding tank. The mussel also can be placed in a container of clean water for several minutes (up to an hour) to collect any additional larvae that may be expelled.

(a)

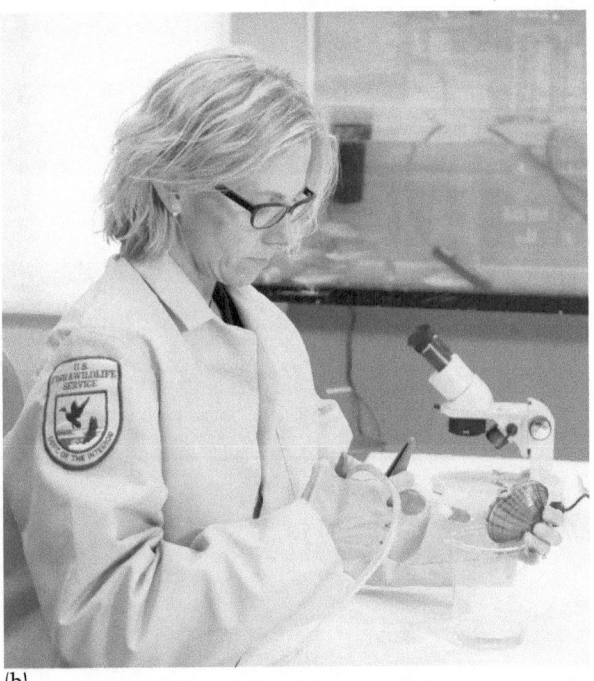

(b)

FIGURE 5.2 (a) A metal bolt can be used to hold the two valves apart while the syringe is inserted. The Petri dish below captures the harvested glochidia. (b) Extracting glochidia from the plain pocketbook (*Lampsilis cardium*) using the syringe method. The mussel is held over a clean container with the ventral margin facing up and the anterior end facing away. (c) The needle is gently inserted into the gill tissue from the posterior end, perpendicular to the water tubes. (d) Glochidia being expelled from the water tubes as water is injected from the syringe. (e) The remaining glochidia being removed from the interior of the shell using a squirt bottle. (f) The flattened gill tissues after the harvest of glochidia is complete.

Photos: Ryan Hagerty, USFWS.

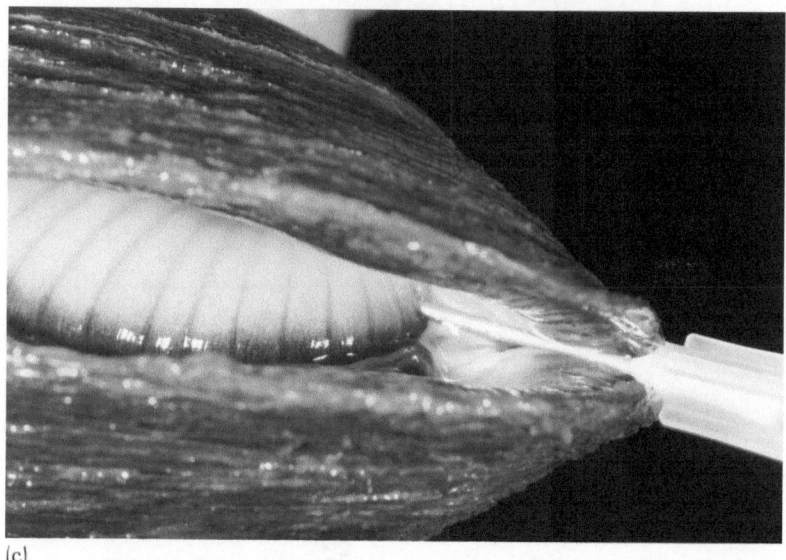

(c)

(d)

FIGURE 5.2 (*cont.*)

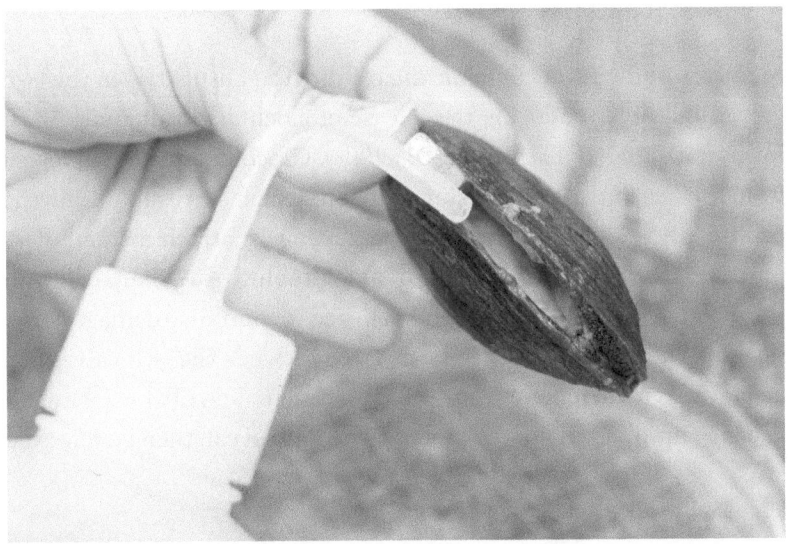

(e)

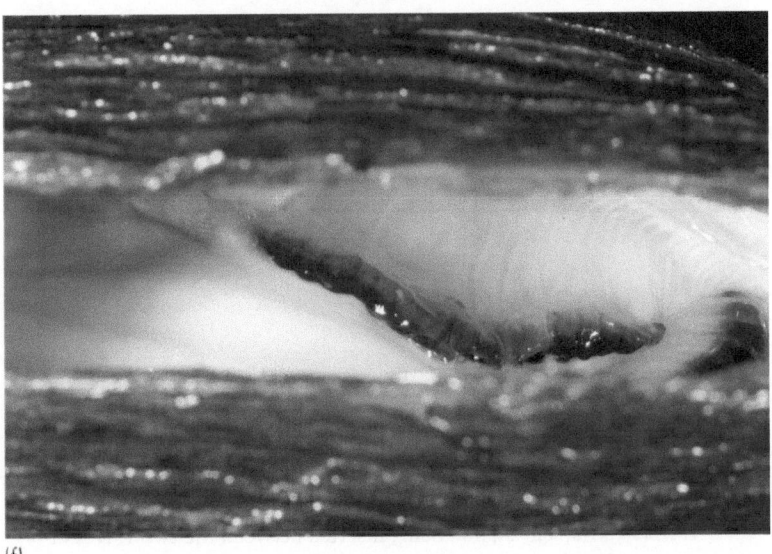

(f)

FIGURE 5.2 (*cont.*)

5.1.2   Siphon Method

Members of the genus *Epioblasma* capture the host between the two valves and hold on during the release of glochidia (Mulcrone, 2004; Jones *et al.*, 2006b; Barnhart *et al.*, 2008). Glochidia can be harvested from species in this genus using a siphon, which serves to mimic the captured host. To create the siphon apparatus, attach a small glass pipette to the end of several feet of nylon tubing. Any sharp edges on the pipette should be blunted and polished to minimize damage to the mussel. The pipette can then be inserted between the two valves of a displaying gravid female (Figure 5.3a). The female will close the shell and grab onto the end of the pipette. The mussel can then be removed

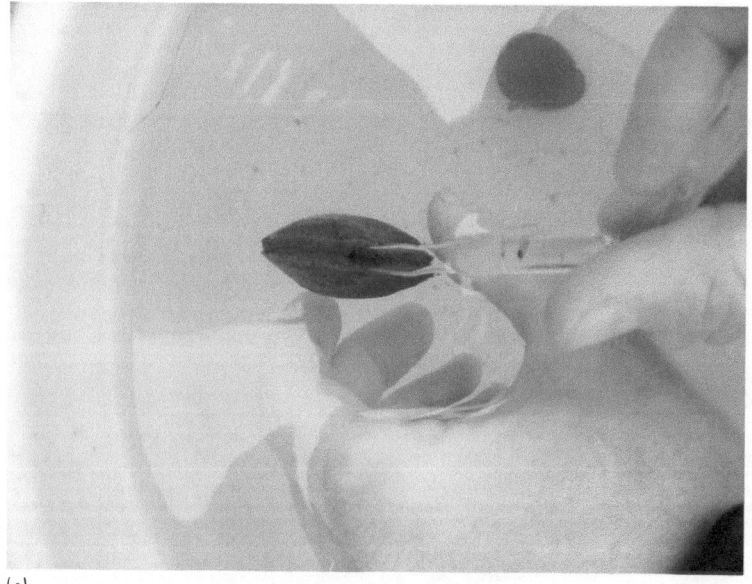

(a)

FIGURE 5.3 (a) The siphon being inserted between the two valves of a gravid female golden riffleshell (*Epioblasma florentina aureola*). (b) Transporting a gravid female to the shore to extract glochidia using the siphon method. The pipette with the mussel attached is held in one hand and the other end of the nylon tube is sealed with the other hand. (c) All of the water collected using the siphon method is passed through a 105 μm mesh screen to capture the glochidia.
Photos: (a) Nathan Eckert, USFWS. (b) Amanda Wood, Virginia Department of Game and Inland Fisheries. (c) Nathan Eckert, USFWS.

(b)

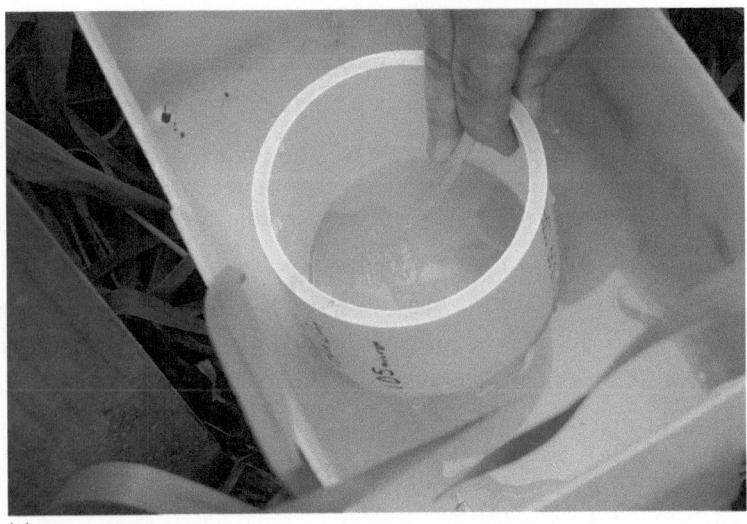

(c)

FIGURE 5.3 (*cont.*)

from the substrate, placed in a container, and transported to shore (Figure 5.3b). Be sure to seal off the opposite end of the nylon tubing to prevent the siphon from losing prime. On shore, place the gravid female in a bucket of water and pass all water exiting the siphon hose through a 105 micron sieve to capture the glochidia (Figure 5.3c). Glochidia are then rinsed from the sieve and either placed into an infestation bath in the field or transported back to the laboratory. If the siphon tube clogs with gill or mantle tissue, the siphon water can be run in reverse to collect glochidia. Because the siphon method is less invasive than the syringe method, it can be very useful for imperiled species. In fact, glochidia can be harvested with the siphon method without removing the gravid female from the stream.

5.1.3 Warming Method

A few long-term brooders produce conglutinates in the fall and hold them over winter (i.e. *Cyprogenia, Dromus,* and *Strophitus*). These species are less sensitive to handling than the conglutinate-producing short-term brooders that tend to prematurely release conglutinates. They can be held in chilled water for months under refrigeration without releasing conglutinates. In the wild, the conglutinates are released in the spring as water temperatures rise. This spring warming can be mimicked in the laboratory by allowing the mussels to warm gradually to room temperature overnight (Figure 5.4). Most conglutinate-producing long-term brooders will release at least a portion of their brood within hours of reaching room temperature. If necessary, the mussels can be held at room temperature for a couple of days to facilitate release. If the warming method does not initiate a timely release of conglutinates, chill the mussels and repeat the warming process. Chemical inducement (Section 5.1.4) also can be used to speed the release of conglutinates if the warming method is ineffective.

5.1.4 Chemical Inducement Method

Some long-term brooders (e.g. *Ptychobranchus*) release small numbers of conglutinates at a time over an extended period. It may not be

FIGURE 5.4 Extracting conglutinates from the James spinymussel (*Pleurobema collina*) using the warming method. The white conglutinates can been seen in the lower left of the photograph. Photo: Rachel Mair, USFWS.

practical to wait for the entire brood to be released and completing multiple small host infestations can be time-consuming. In this case, serotonin can be used to induce gravid females to release all conglutinates at once (Figure 5.5). A bath of 20–40 ppm seratonin for 4–6 hours is generally sufficient to induce release. Serotonin creatinine sulfate monohydrate is the preferred source of commercially available serotonin, but be aware this compound is not 100% serotonin. Be sure to use the percent active ingredient on the label when calculating the proper treatment concentration. Serotonin acts as a muscle relaxer and can slow the closing speed of the glochidia. Remove all conglutinates from the serotonin bath as quickly as possible and place them in fresh water to avoid poor attachment to the host during infestation.

(a)

(b)

FIGURE 5.5 (a) Female brook floater (*Alasmidonta varicosa*) prior to exposure to serotonin to harvest glochidia. (b) Female brook floater (*Alasmidonta varicosa*) 1 hour post-exposure to serotonin.
Photo: Rachael Hoch, North Carolina Wildlife Resources Commission.

The chemical inducement method is still relatively experimental, so test the method on small batches of mussels and take detailed notes to help improve future results. In the United States, exposing federally listed species to serotonin or any other chemicals to induce conglutinate release requires a permit.

5.2 HARVESTING LARVAE: SHORT-TERM BROODERS

Short-term brooders typically are more challenging to propagate than long-term brooders. The window of opportunity to collect gravid females from the wild is short (weeks or even days) and tends to occur in the spring when high water levels can make collection of the gravid females difficult. Short-term brooders also are more sensitive to handling, temperature changes, and other stressors. If gravid females are collected too early, handling stress may result in the premature release of eggs or immature glochidia. If gravid females are collected too late, the glochidia may have already been released for the year. In both cases, propagation for that species will have to wait until the following year.

In most short-term brooders, the larvae are not held loose in the water tubes. Instead, the glochidia are packaged in conglutinates. Attempting to harvest conglutinates using the syringe method can do extensive damage to the gill tissues. Less invasive methods of harvesting glochidia from short-term brooders include the isolation method and the suction method.

5.2.1 Isolation Method

The isolation method must begin with proper handling of gravid females in the field. If a short-term brooder is exhibiting visual cues of larval brooding in the wild (i.e. mantle display), the female mussel should be placed directly into a sealed plastic bag underwater to capture any conglutinates released in response to handling. If a display is not visible, minimize the length of time mussels spend in a collection bag and place them in a sealed plastic bag as soon as gravidity can be confirmed. The water in the sealed plastic bags should be exchanged

FIGURE 5.6 Conglutinates of the sheepnose (*Plethobasus cyphyus*) collected using the isolation method. The white conglutinates show up nicely against the black holding container.
Photo: Nathan Eckert, USFWS.

regularly to maintain dissolved oxygen, but be careful not to spill any conglutinates that may have been released in the bag.

At the laboratory, gravid females should be placed in a container filled with water from the collection site. To aid in the retrieval of expelled conglutinates, remove all substrate from the container and use a dark-colored container to make the light-colored conglutinates easier to see (Figure 5.6). Light aeration can be used to maintain dissolved oxygen concentrations in the container, but excess aeration can jostle the conglutinates, leading to premature closure of the glochidia. A partial water change every 24 hours will keep the mussels submerged and limit disturbance, while minimizing the build-up of ammonia and other waste products. Most short-term brooders will release conglutinates within 24–48 hours. Expelled conglutinates can be collected with a large pipette or all water in the container can be poured through a 105 micron sieve.

5.2.2 Suction Method

The suction method is used to harvest glochidia from mussel species that transfer conglutinates to a mantle magazine prior to release

(Section 2.8.3). Members of the genus *Quadrula* utilize a mantle magazine to temporarily hold conglutinates just inside the excurrent aperture (Barnhart *et al.*, 2008). To harvest glochidia using the suction method, insert a turkey baster or large transfer pipette about one half inch inside the mantle magazine and draw water out (Figure 5.7). A baster or pipette with a clear shaft works best because the conglutinates can easily be seen exiting the mussel. Immediately transfer any harvested conglutinates to a clean beaker of water. In most cases, a single draw of water will suffice. If the gravid female does not retract the mantle magazine within a few minutes, a second suction may be needed to remove additional glochidia.

5.3   PREPARING CONGLUTINATES FOR INFESTATION

In most cases, the individual glochidia must be separated from a conglutinate prior to checking viability or infesting the host. Two techniques for dislodging glochidia from a conglutinate will be described. For pelagic and mucoid conglutinates (Section 2.8.2), place the conglutinates on a set of nested sieves and spray with water. For most species, use a 300–400 micron mesh sieve on top to catch the conglutinate packaging material and a 100–150 micron mesh sieve on the bottom to catch the glochidia. For conglutinates with a solid core (i.e. demersal and superconglutinates), more force is required to dislodge the glochidia. In this case, draw the conglutinates into the head end of a plastic pipette or turkey baster and shake the pipette or tap it against a counter top (Figure 5.8). After several seconds of shaking, invert the pipette and release the water and glochidia into a clean container. Repeat as needed until all glochidia are completely stripped from the conglutinates.

5.4   CHECKING THE VIABILITY OF GLOCHIDIA

After harvest from the female, it is very important to check the viability of the glochidia. Infesting a host with glochidia that are immature or non-viable is likely to fail. To test the viability of glochidia, place a subsample of glochidia (100 or less) in a Petri dish and add

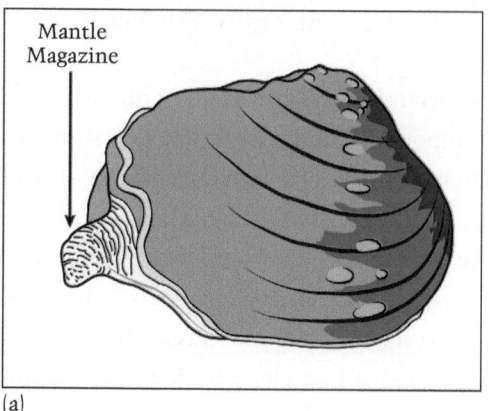

(a)

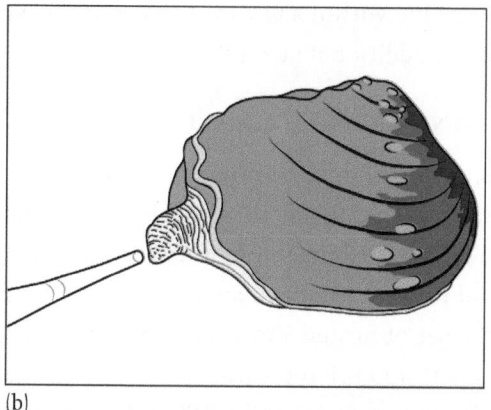

(b)

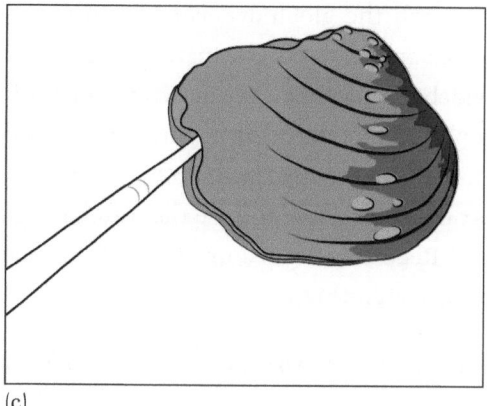

(c)

FIGURE 5.7 (a) Gravid female winged mapleleaf (*Quadrula fragosa*) with the mantle magazine protruding from the posterior shell margin. (b) A transfer pipette being inserted into the mantle magazine to extract conglutinates using the suction method. (c) Once the conglutinates are removed, the female withdraws the mantle magazine inside the shell. Graphics: Kristin Simanek, USFWS.

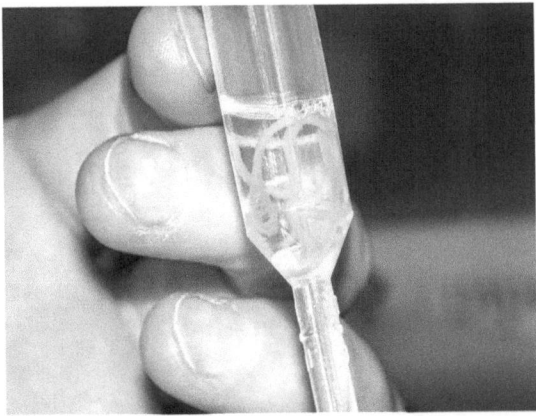

FIGURE 5.8 Conglutinates of the fanshell, *Cyprogenia stegaria*, drawn into a transfer pipette to dislodge the glochidia.
Photo: Chris Barnhart, Missouri State University.

a few grains of salt (Zale and Neves, 1982). As the salt dissolves in the water, monitor the glochidia under a dissecting microscope. The typical response to the presence of salt is contraction of the larval adductor muscle and abrupt closure of the valves. While the speed of the valve closure may be relatively slow in cold water, valve closure should begin within a few seconds of adding salt. When all of the salt grains have dissolved, agitate the water in the Petri dish to thoroughly mix the solution. Count all open (Figure 5.9a) and closed (Figure 5.9b) glochidia, and calculate percent viability using the following formula:

$$\% \text{ viability} = \frac{\text{number of closed glochidia}}{\text{number of closed glochidia} + \text{number of open glochidia}} \times 100\%$$

If viability is greater than 90%, proceed with the infestation. If viability is less than 90%, re-test another subsample of 100 glochidia. If viability in the second test is below 90%, consider either discarding the glochidia or increasing the number of glochidia added to the infestation bath. If viability is less than 90%, but the mussel species is highly imperiled, it may be crucial to still proceed with

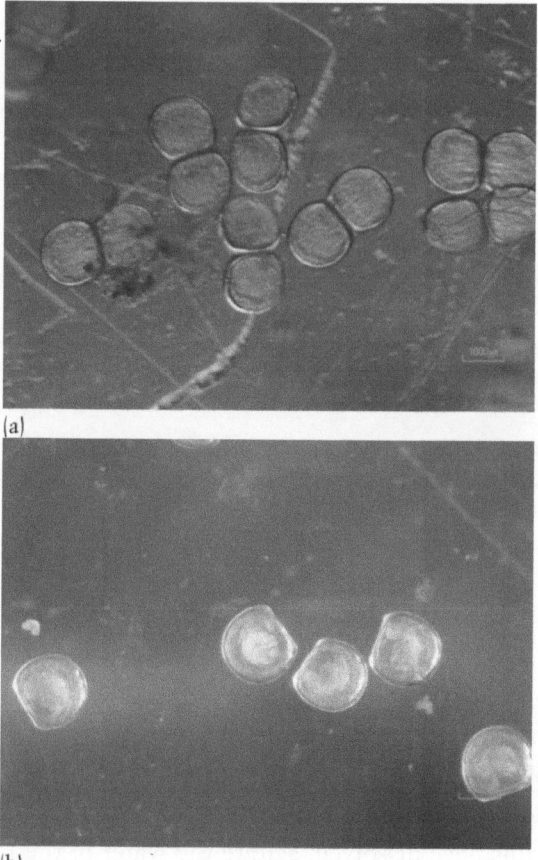

FIGURE 5.9 (a) Glochidia of the golden riffleshell (*Epioblasma florentina aureola*) immediately following extraction from the gravid female. The glochidia are gaping wide open awaiting attachment to the host. (b) Glochidia of the golden riffleshell (*Epioblasma florentina aureola*) after the addition of a small amount of salt to the Petri dish to test viability. The glochidia are closed, indicating they are viable or capable of attaching to the host.
Photos: Megan Bradley, USFWS.

the infestation. The number of juveniles recovered may be low and juvenile growth and survival may be compromised, but the high level of imperilment makes it worth the risk. In most cases, viability will be near 100%.

The salt test also can be a good indicator of mussel infectivity, but only when viability is 90% or more (Fritts *et al.*, 2014). Viabilities under 90% resulted in low rates of infectivity, especially for glochidia harvested near the end of the brooding cycle. Keep in mind the salt test does not guarantee the glochidia are mature and in good condition.

5.5 ESTIMATING THE NUMBER OF GLOCHIDIA

If the glochidia are viable, the next step is to estimate the total number of glochidia. While this step is not required, an estimate of the total number of glochidia will help: (1) prevent over-infestation and potential death of the host, (2) use glochidia more efficiently, which can be a concern when working with imperiled species or species with low fecundity, (3) determine the optimal number of glochidia needed to facilitate rapid attachment, and (4) calculate the rate of glochidia attachment.

To estimate the total number of glochidia, suspend all the glochidia from a single gravid female in a graduated beaker (1 L or smaller) and record the total water volume in milliliters. Using a turkey baster, draw water in and out of the beaker to create a uniform suspension of glochidia. The water depth in the beaker should be higher than the width of the beaker to facilitate uniform mixing. While the glochidia are being mixed, use the opposite hand to withdraw ten, 200 microliter subsamples using a volumetric pipette (Figure 5.10a). Place each subsample on a dry Petri dish (Figure 5.10b). Count the number of glochidia in each subsample under a dissecting microscope. The number of glochidia per milliliter can be calculated using the following formula (the volume of sample in this case is 2 mL (10 × 200 microliter samples)):

$$\text{Number of glochidia per mL} = \frac{\text{number of glochidia}}{\text{volume of sample (mL)}}$$

The total number of glochidia can be calculated using the following formula:

Total number of glochidia = glochidia per mL × total mL in the beaker

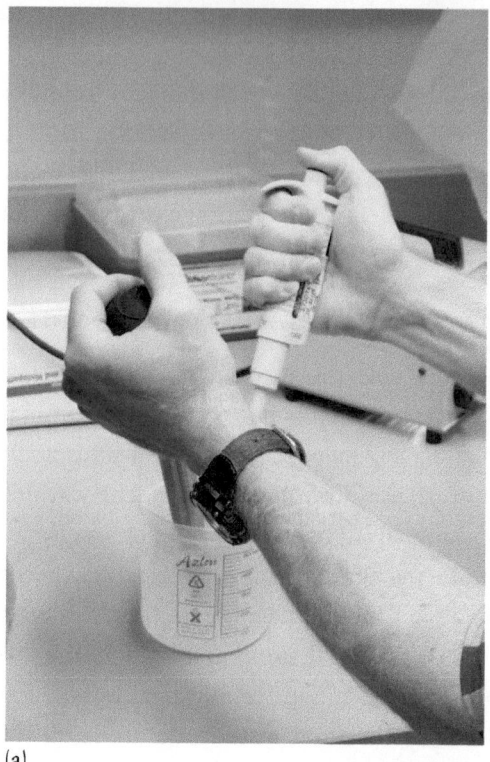

(a)

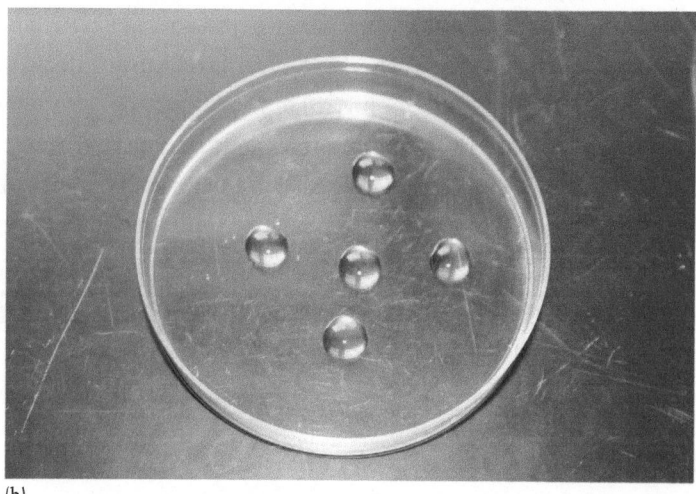

(b)

FIGURE 5.10 (a) Glochidia being mixed with a turkey baster while 10 × 200 microliter subsamples are collected for enumeration. (b) The 200 microliter subsamples are placed on a dry Petri dish and the number of glochidia in each drop are counted.
Photos: Matthew Patterson, USFWS.

For ease of sampling, the subsample volume (200 microliters) can be adjusted to capture 20–30 glochidia per droplet. Larger numbers of glochidia per subsample can be difficult to count, and increase the variability of the estimate. The water volume in the beaker also can be adjusted to achieve the ideal density of glochidia per subsample. Use 1–2 L for highly fecund species, 500 ml for moderately fecund species, and 200 ml for small mussel species with low fecundity.

5.6 METAMORPHOSIS

Survival and fitness of the larvae begin to decline as soon as they are harvested from the adult female, and the rate of decline seems to be more rapid in the short-term brooders (Fritts *et al.*, 2014). Consequently, metamorphosis should begin as soon as possible after harvest. Metamorphosis can be completed either by attaching the glochidia to a suitable host (*in vivo*) or placing the glochidia in a suitable culture medium (*in vitro*).

5.6.1 *In Vivo Metamorphosis*

In the literature, attaching larval freshwater mussels to a host in the laboratory has been called both an "infestation" and an "inoculation". An infestation generally refers to the introduction of parasites, while an inoculation refers to the introduction of an agent that stimulates the production of antibodies. Both terms accurately describe the process and will be used interchangeably in this chapter.

5.6.2 *Infestation Bath*

The most widely used method for attaching mussel larvae to a host is the infestation bath. The harvested larvae and a suitable host are placed together in a container to facilitate attachment. Selecting the appropriate container size and water level for the infestation bath are very important (Figure 5.11). The host can easily become stressed if the container is too small (either for the number of hosts or the size of the host species) or the water volume is too small. If the host dies, either during or after the infestation process, the larvae will be

(a)

(b)

FIGURE 5.11 (a) An 18 gallon plastic tub with rope handle being used for an infestation bath. (b) A two gallon bucket being used for an infestation bath. (c) Any small plastic container that will hold water can be used for an infestation bath. (d) Darters being held in a 10 L Aquatic Habitat tank for infestation. Enough water is added to just cover the backs of the fish host.
Photos: Nathan Eckert, USFWS. (A black-and-white version of part (d) of this figure will appear in some formats. For the color version, please refer to the plate section.)

(c)

(d)

FIGURE 5.11 (cont.)

Table 5.2. *Commonly used infestation containers and water levels for various host species*

Container	Species	Host number	Water level	Infestation time[a]
3L AHAB tank	darters, sculpin	10–20	1″	45–60 min
10L AHAB tank	darters, sculpin	50–100	1″	45–60 min
5 gallon bucket	bass, sunfish	5–15	1 gallon	10–35 min
17 gallon rope handle tub	any species	50–200	9 gallon	Varies with species

[a]Infestation time is highly variable and over-infestation can kill the host. Use caution and check the fish gills regularly throughout the infestation process.

unable to complete metamorphosis. If the container or water volume is too large, the chances of the larvae coming in contact with the host decreases, increasing the time needed to achieve the desired infestation rate. The ideal container size and water volume will depend upon the host species and number of hosts used. Small hosts like darters and sculpin will tolerate small containers and very small water volumes. If the fish can position themselves normally in the water without breaking the water surface, the water volume is sufficient (Figure 5.11d). Black bass, walleye, and catfish require larger containers and water volumes (>1 L per fish). Table 5.2 provides some basic guidelines for container size and water volume for various host species.

All containers selected for the infestation bath should be labeled "infestation only" (Figure 5.12a). Proper labeling will help ensure infestation containers are not used for chemical treatments or other hatchery procedures that may be toxic to mussel larvae.

Once a container has been selected, the infestation bath is filled with water that is the correct temperature and free of fish mucus. The infestation bath temperature should be within a few degrees of the water temperature in the system used to hold the host prior to

(a)

FIGURE 5.12 (a) An infestation bath container labeled infestation only.
(b) Largemouth bass (*Micropterus salmoides*) being poured through a
net to remove all fish mucus before transfer to the infestation bath.
(c) Largemouth bass (*Micropterus salmoides*) being transferred from
the net to the infestation bath. Rachel Mair (USFWS) is holding a
Petri fish full of glochidia and will pour them into the infestation
bath immediately following the fish. (d) Glochidia are immediately
poured into the infestation bath with aeration. (e) Largemouth bass
(*Micropterus salmoides*) in the infestation bath with the airstone
removed to show the density of fish.
Photos: Ryan Hagerty, USFWS.

(b)

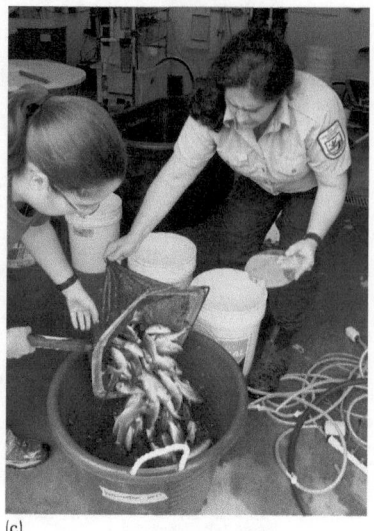

(c)

(d)

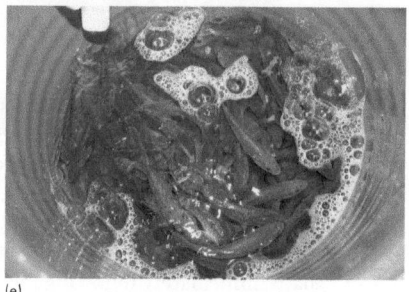

(e)

FIGURE 5.12 (cont.)

and after infestation. Do not match water temperatures by simply using water from the host holding system. Larvae can prematurely close on fish scales, mucus or other bodily fluids present in the water before attaching to the host. Larvae also may close prematurely if the holding system was recently treated with salt to limit stress on the host. Instead, the infestation bath should be filled with water that has never held fish (spring water, well water, etc.).

Pour the hosts through a net to drain off any water and fish mucus before adding them to the infestation bath (Figure 5.12b). The host should then be transferred from the net to the infestation bath, followed immediately by the larvae (Figure 5.12c, d).

High densities of host fish and thousands of parasitic larvae can make conditions in the infestation bath stressful (Figure 5.12e). Dissolved oxygen concentrations can drop quickly. To maintain dissolved oxygen, an air stone should be added. Aeration has the added benefit of keeping the larvae in suspension, increasing the chances of attachment. The intensity of aeration should be adjusted based on the attachment strategy of the larvae. Glochidia that attach to the gills of the host should be suspended with vigorous aeration. Glochidia that attach to fins or other soft tissues typically attach better if aeration is kept to a minimum and water levels are shallow. This allows the glochidia to settle to the bottom of the infestation bath and come in close contact with the host (Mark Hove, UMN, personal communication).

The optimal number of larvae to add to the infestation bath depends on the propagation goal, the host species, and the mussel species. Infestations aimed at large-scale production of juveniles for restoration and recovery are typically handled differently from research infestations. An infestation for production will attempt to attach as many larvae as possible without killing the host. Smaller fish, such as darters and minnows, held in shallow water can be effectively infested with about 1000 glochidia per liter. When infesting larger fish species like smallmouth bass (*Micropterus dolomieu*) and largemouth bass (*Micropterus salmoides*) with plain pocketbook (*Lampsilis cardium*), fatmucket (*Lampsilis siliquoidea*), pocketbook (*Lampsilis ovata*), wavy-rayed lampmussel (*Lampsilis fasciola*) or the Higgins eye pearly mussel (*Lampsilis higginsii*), 4000 to 5000 glochidia per liter should be added to the infestation bath. Experiment with each mussel–host pairing to determine the optimal water level, larval abundance, larval density, and host density in the infestation bath. For research infestations (host species testing or toxicity studies), it is not necessary to attach a maximum number of glochidia to each host. Dozens to a few hundred glochidia per fish will yield the required results. A small infestation bath with approximately 1000 glochidia per fish will yield approximately 100–300 glochidia attached per fish in about 15 minutes. The percentage of glochidia added to the infestation bath that actually attach to the host varies with species and conditions in the infestation bath. Barnhart (2004) found the following percentage attachment rates for several mussel–fish-host combinations: 36–56% for pink mucket (*Lampsilis abrupta*) on largemouth bass (*Micropterus salmoides*), 45% for the black sandshell (*Ligumia recta*) on walleye (*Sander vitreus*), and 57–61% for the Neosho mucket (*Lampsilis rafinesqueana*) on largemouth bass (*Micropterus salmoides*).

Over-infestation has the potential to kill the host, so all fish should be monitored closely for signs of stress throughout the infestation process. Gasping at the surface, erratic behavior, or loss of equilibrium may indicate a drop in dissolved oxygen or a near lethal dose of glochidia. The host species should be removed from the infestation bath and returned to the holding tank if any of these signs of stress are observed. Even if no obvious signs of stress are observed, it is still important to monitor the host regularly to assess the extent of infestation or the number of larvae attached. Some host species can tolerate heavier loads of glochidia than others. Largemouth bass (*Micropterus salmoides*) and smallmouth bass (*Micropterus dolomieu*) can tolerate heavy glochidia loads and rarely die from over-infestation. Darters, minnows, and freshwater drum (*Aplodinotus grunniens*) on the other hand may show signs of stress even at low levels of infestation. To assess the extent of infestation, remove a couple of fish from the infestation bath every 5–10 minutes. For hooked glochidia that attach to external surfaces, examine the fins to look for attached glochidia. For hookless glochidia that attach to gill tissues, gently pull back the operculum and examine the gills to determine the extent of the infestation (Figure 5.13a). To the naked eye, the glochidia will appear as small white specks on the bright red gill tissues (Figure 5.13b). Quantifying the extent of infestation in this manner can be difficult with smaller fish species. An alternative method is to sacrifice two or three fish from the infestation bath, excise the gills, and examine them under a microscope. For species that can tolerate heavier loads of glochidia, the infestation can continue until the outer edge of the gills is completely white (Figure 5.13c). For less tolerant species, stop the infestation once the outer gill edge first begins to turn white. If a host species is sensitive to handling, a mild anesthetic can be used to decrease stress after infestation. Anesthetic will likely cause glochidia to close prematurely so it should not be used in the infestation bath.

While the host is being monitored for signs of stress and extent of infestation, the infestation bath also should be monitored for the percentage of glochidia that remain open. If the majority of the glochidia that remain in suspension are closed and no longer able to attach to the host, the infestation process should be stopped to limit unnecessary stress. Remove a small volume of water from the infestation bath every 10–15 minutes and examine it under a dissecting microscope. When the percentage of open glochidia drops below 30%, stop the infestation and return the fish to the holding system.

It may be necessary to experiment with each mussel–host species pairing to ensure optimal infestation. During the experimental phase, keep a close eye on the fish and take detailed notes on the number of glochidia per liter, the number of host fish, the container size, the water volume, the duration of the infestation process, etc. The lessons learned will help improve future infestations. Table 5.3 provides some infestation protocols for a variety of freshwater mussel species propagated at the Virginia Department of Game and Inland Fisheries' Aquatic Wildlife Conservation Center.

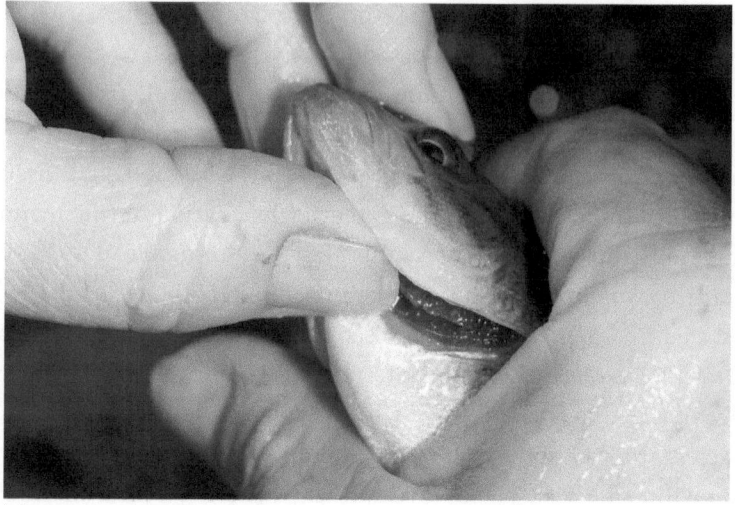

(a)

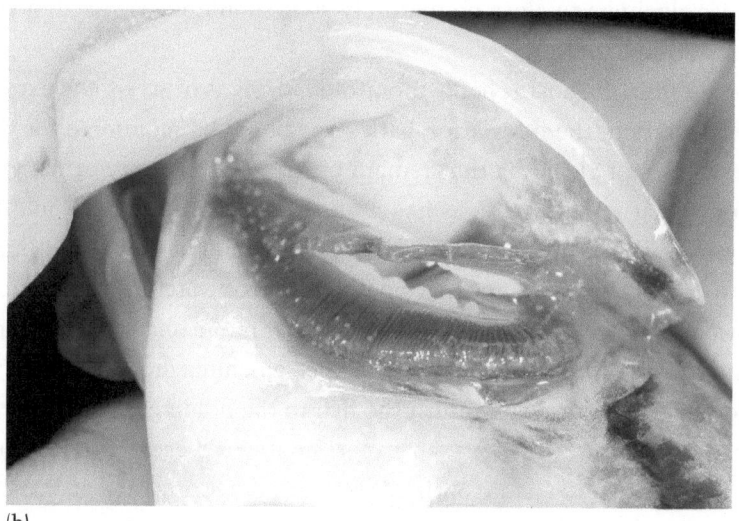

(b)

FIGURE 5.13 (a) To monitor the gills of the fish host during the infestation process, hold the fish gently, pull back the operculum, and check for attached glochidia. (b) The gills of the fish host after 5 minutes in the infestation bath. The small white specks on the gill are attached glochidia. (c) The gills of the fish host after 30 minutes in the infestation bath. The number of attached glochidia has increased significantly, forming a white ring around the edge of the gill. Photos: Ryan Hagerty, USFWS. (A black-and-white version of parts (b) and (c) of this figure will appear in some formats. For the color version, please refer to the plate section.)

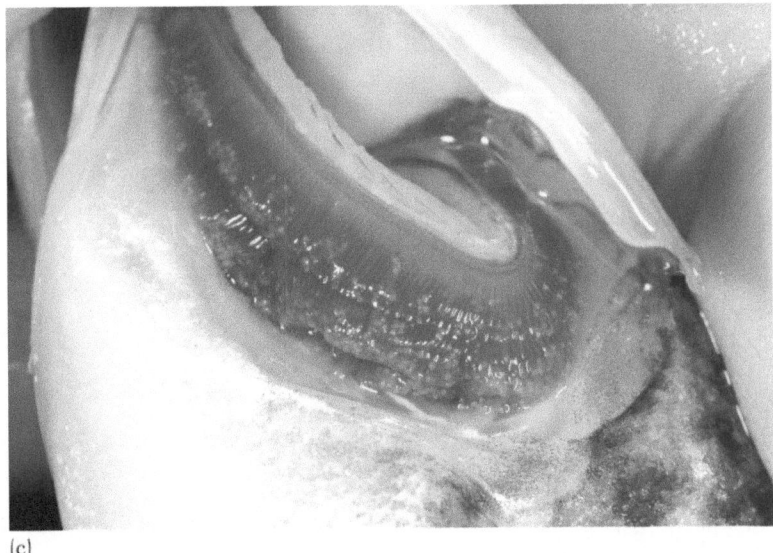

(c)

FIGURE 5.13 (cont.)

5.6.3 Direct Pipette Infestation

Larvae also can be placed directly onto the gills of the host using a pipette (Figure 5.14a). Larvae can be pipetted onto the gills either through the mouth or by gently lifting the operculum. The host should be anesthetized during this process. Direct pipette infestations can be advantageous when either a small number of glochidia or a small number of host fish are available. The process can be very time-consuming so infesting large numbers of fish is not practical. Care also must be taken to ensure the host is not over-infested.

Direct pipetting is the preferred method for infesting mudpuppies (*Necturus maculosus*) with the salamander mussel (*Simpsonaias ambigua*). Place the mudpuppies in a shallow dish with water just deep enough to cover the gills (Figure 5.14b). Pipette the glochidia directly onto the gills, which are located externally just behind the head (Figure 5.14c). Larval threads in the salamander mussel cause the glochidia to stick together in clumps. The glochidia can be re-used until the clumps begin to disassociate.

Table 5.3. *Infestation protocols for freshwater mussels propagated at the Virginia Department of Game and Inland Fisheries' Aquatic Wildlife Conservation Center (Marion, VA)*

Genus/species	Timing	Harvest method	Host fish	Fish/ infestation	Mussels/ infestation	Container/ volume	Infestation time (min.)	Days to drop-off	Instructions	Expected # juveniles
Actinonaias	Mar–May	Syringe	LMB/SMB	50–200	1 mussel	Large tub 9 gallons	15–45	10–16	Use 1 gill for 50–100 4–6″ LMB. Infest only 50 fish per tub; repeat	20 000–65 000
Cyprogenia stegaria	Mar–Apr	Warming Chemical Isolation	Roanoke darter Logperch	Based on glochidia	All available	9L AHAB 1–1.5″	45–60	14–21	5 darters or 1 large logperch per conglutinate	500–5000
Dromus dromas	Feb–Apr	Warming Chemical Isolation	Logperch Guilt darter	50	1 mussel or less	9L AHAB 1–1.5″	30–60	13–16	1 to 2 conglutinates per fish	4000–7000

Elliptio dilatata	Jun–Jul	Isolation Needle	Sculpin	80–95	1 mussel	9L AHAB 1.5–2″	30–45	8–10	Add extra fish if using more than 1 mussel.u	5000–10 000
Epioblasma brevidens	Mar–Jun	Syringe Siphon	Sculpin Logperch	80	1 mussel	9L AHAB 1.5–2″	60	10–14	Use less water if using two tanks. Crowd fish after 40 minutes if needed	4000–9000
Epioblasma capsaeformis	Mar–Jun	Syringe Siphon	Sculpin Redline darter	80	2–3 mussels	9L AHAB 1.5–2″	60	8–10	Use 1 or 2, 9L tanks. Crowd fish after 40 min if needed.	5000–12 000
Epioblasma triquetra	Apr–Jul	Syringe Siphon	Sculpin Logperch	80	2 mussels	9L AHAB 1.5–2″	60	10–12	Crowd fish after 40 minutes if necessary	3000–5000

(cont.)

Table 5.3. (cont.)

Genus/species	Timing	Harvest method	Host fish	Fish/infestation infestation	Mussels/infestation	Container/volume	Infestation time (min.)	Days to drop-off	Instructions	Expected # juveniles
Epioblasma f. walkeri	Apr–May	Siphon Syringe	Sculpin Fantail darter	80	1 mussel	9L AHAB 1–1.5"	60	8–10	Normally a streamside infestation. Do not use syringe unless absolutely necessary	1500–4500
Fusconaia cor	May–Jun	Isolation	Warpaint shiner Striped shiner Whitetail shiner	All available	1 mussel	5 gal. bucket 2 gallons	45–60	9–14	Infest in a bucket or feed whole conglutinates	????
Lampsilis fasciola	Mar–Oct	Syringe	LMB/SMB	50	1 mussel	Large tub 9 gallons	30–50	9–14	Crowd fish for 10 000–30 minutes. Use 1 gill or re-infest following day if necessary	10 000–50 000

Lampsilis ovata	Mar–Oct	Syringe	LMB/SMB	50–100	1 gill	Large tub 9 gallons	30–60	9–14	One mussel can infest up to 200 fish. Same procedure as *L. fasciola*	10 000–50 000
Lasmigona costata	Apr–May	Isolation	Sculpin	80	1 mussel	9L AHAB 2–3″	45–60	7–10	Do not crowd fish to avoid over-infestation	2000–5000
Lasmigona holstonia	Apr–May	Isolation	Sculpin	80	1 mussel	9L AHAB 2–3″	45–60	7–10	Glochidia can attach to fins, crowding isn't needed	7000–15 000
Lemiox rimosus	Mar–May	Syringe	Greenside darter Snubnose darter	20–30	1 mussel	9L AHAB 1–1.5″	60	14–30	Crowd fish by tilting tank up on one end	1000–1500

(cont.)

Table 5.3. (cont.)

Genus/species	Timing	Harvest method	Host fish	Fish/ infestation	Mussels/ infestation	Container/ volume	Infestation time (min.)	Days to drop-off	Instructions	Expected # juveniles
Ligumia recta	Apr–Jul	Syringe	Walleye	50–100	1 mussel	Large tub 9 gallons	30–60	10–14	Prefer hatchery walleye. If wild, use as many as possible	5000– 25000
Ptychobranchus	Mar–May	Warming Chemical	Redline darter	50	100 conglutinates	9L AHAB 1–1.5″	30–45	12–14	Use 2 conglutinates per fish. Do not exceed 100 in an infestation bath. Repeat bath for extra fish	3000– 10000

Species	Season	Method	Host fish		Dose	Container	Temp	Days	Notes	Glochidia
Villosa iris	Mar–Oct	Syringe	Rock bass	10–15	1 mussel	5 gal. bucket 2 gallons	30–60	10–14	Double or triple glochidia if using Copper Creek mussels	5000–16000
Villosa perpurpurea	Mar–Apr	Syringe	Sculpin Fantail darter	80	1 mussel	9L AHAB 1–1.5"	60	8–14	Crowd fish by tilting tank up on one end	1000–3000
Villosa vanuxemensis	Mar–Aug	Syringe	Sculpin	160	1 mussel	9L AHAB 1.5–2"	60	9–14	Half infestations can be done if fish are limiting	4000–13 000

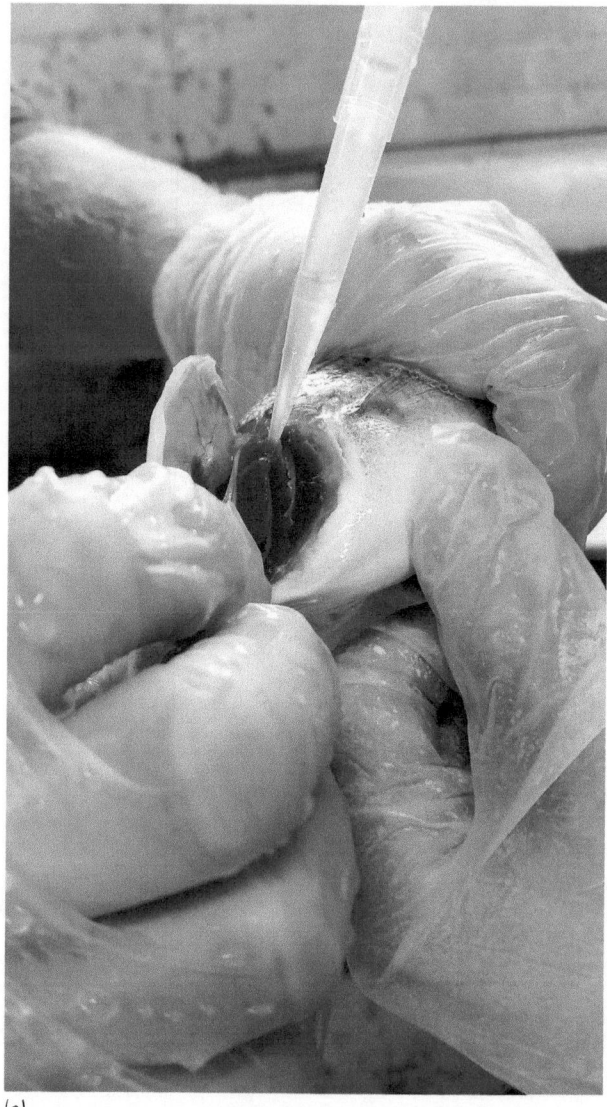

(a)

FIGURE 5.14 (a) Pipetting glochidia directly onto the gills of an anesthetized fish. (b) Mudpuppies are typically placed in a shallow container when infesting with the salamander mussel (*Simpsonaias ambigua*). (c) Glochidia are pipetted directly onto the gills of the mudpuppy using a plastic transfer pipette.

Photos: (a) Amy Maynard, Conservation Management Institute, Virginia Polytechnic Institute and State University. (b, c) Megan Bradley, USFWS.

(b)

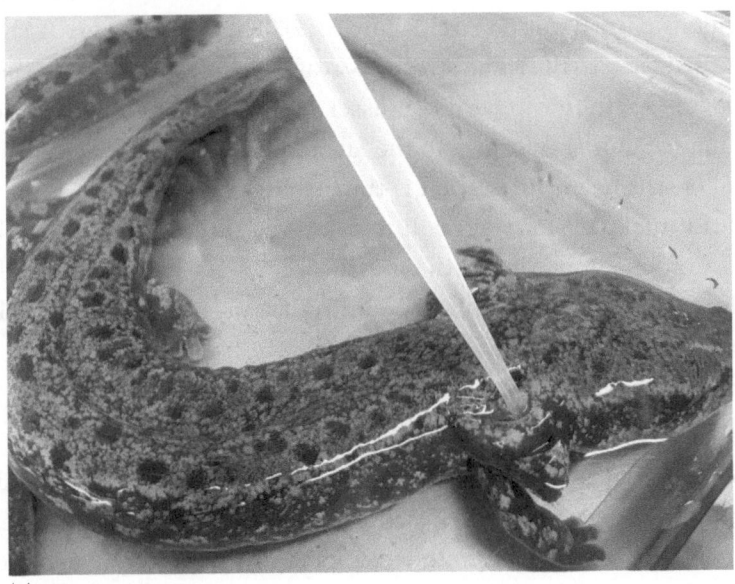

(c)

FIGURE 5.14 (cont.)

5.6.4 Natural Infestation

A host can be "naturally" infested with larvae using a couple of techniques. First, a gravid female and a suitable host can be placed in the same holding system, allowing the infestation to occur as it would in the wild. For this technique to be effective, the gravid female and the host need to be in close proximity to increase the chances of an encounter. A small- to medium-sized holding system (e.g. 10 gallon aquarium) is preferred.

Second, the host can be hand-fed conglutinates harvested from the gravid female. Most conglutinates are designed to mimic prey items so the hosts are typically willing to eat them. Separate the hosts into small groups and hand-feed each group one at a time. If all the hosts are hand-fed at once, it can be nearly impossible to separate infested from uninfested fish after the infestation process is complete.

5.6.5 Streamside Infestation

A suitable host also can be infested in the field. Gravid females and the host species are collected from the same stream and the infestation is carried out on-site (Figure 5.15). The infested host is then released back to the same stream. Streamside infestation can be particularly useful for imperiled mussel species that could be stressed or injured during transport to the laboratory. The primary disadvantage of streamside infestation is the inability to tag mussels prior to release, making it very difficult to assess the success or failure of the propagation effort. Unless the mussel species is extirpated from the release site, it is impossible to confirm that any new recruits are the result of the streamside infestation.

5.6.6 Free Release

Free release is similar to streamside infestation with a minor modification. The host is infested with glochidia in the laboratory, held in captivity until the juveniles are almost ready to drop off, and then transported back to the wild for release. Free release provides a couple

FIGURE 5.15 A streamside infestation using a portable aerator and 10 L AHAB tanks.
Photo: Amanda Wood, Virginia Department of Game and Inland Fisheries.

of advantages over streamside infestation. The infested hosts can be monitored for health issues prior to release and they can be released in the target restoration area just prior to juvenile drop-off. With streamside infestation, the host is typically released weeks prior to juvenile drop-off. During that period, infested fish could leave the target restoration area or even die before metamorphosis is complete. If this infestation technique will be used, be sure to obtain proper permission from the local natural resource agency before releasing fish into the wild.

5.7   MAINTENANCE OF HOST SPECIES POST-INFESTATION

Once the larvae are encapsulated on the host, a variety of both recirculating and flow-through aquaculture systems are available for maintaining the health and survival of hosts in the laboratory.

5.7.1 Recirculating Aquaculture Systems

Multi-tank recirculating systems were described in detail in Chapter 3 (Figure 5.16). The multi-tank system is ideal for host species testing and holding small fish species for juvenile production. The small tank size in the multi-tank system makes it difficult to hold larger fish species like largemouth bass (*Micropterus salmoides*) and walleye (*Sander vitreus*). If the biofilter is not maintained properly or the system is over-loaded with excess biomass, ammonia spikes and other water quality issues can cause mortality. Water quality should be monitored on a daily basis to ensure no negative impacts to the infested fish.

A commonly used large-scale system is the recirculating propagation system or RPS (Barnhart, 2003). The RPS includes one

FIGURE 5.16 A customized multi-tank rack system at the Virginia Department of Game and Inland Fisheries' Aquatic Wildlife Conservation Center outfitted with 10 L tanks for larger-scale production.
Photo: Matthew Patterson, USFWS. (A black-and-white version of this figure will appear in some formats. For the color version, please refer to the plate section.)

or more 250 gallon, conical bottom tanks outfitted with a double standpipe, a sump, mechanical filtration, biological filtration, and UV sterilization (Figure 5.17a, b, c). The inner standpipe controls the water level in the tank while the outer standpipe has slits in the bottom that create water currents along the bottom of the tank

(a)

FIGURE 5.17 (a) Large recirculating propagation system (RPS; Barnhart, 2003) at the White Sulphur Springs National Fish Hatchery (White Sulphur Springs, West Virginia). Two, 250 gallon tanks drain to a central sump. (b) The 250 gallon tank with double standpipe is capable of holding a large number of infested fish. (c) The central sump includes biological filtration, mechanical filtration, and plankton nets attached to the tank outflow to capture juvenile mussels. (d) A view from above an RPS showing the inner and outer standpipes. (e) The slits in the bottom of the outer standpipe create currents along the bottom of the tank that entrain juvenile mussels and carry them to a collection device in the sump. (f) A plankton net hanging over the RPS tank outflow pipe to collect juvenile mussels at the Virginia Fisheries and Aquatic Wildlife Center at Harrison Lake National Fish Hatchery. (g) The plankton nets used in the RPS system at the Virginia Fisheries and Aquatic Wildlife Center at Harrison Lake National Fish Hatchery. The nets are sprayed down to drive any juveniles toward the plastic bottle attached to the bottom of the net. The bottle cap is removed and the juveniles are sprayed into a Petri dish or other container for enumeration. (h) A modified plankton net used at Missouri State University to collect juveniles from the RPS. Two plastic bottles are nested on the end of the net to create a dead space and protect the newly metamorphosed juveniles from the turbulence of the net. Photos: (a, d, e, f, g) Matthew Patterson, USFWS. (b, c, h) Chris Barnhart, Missouri State University.

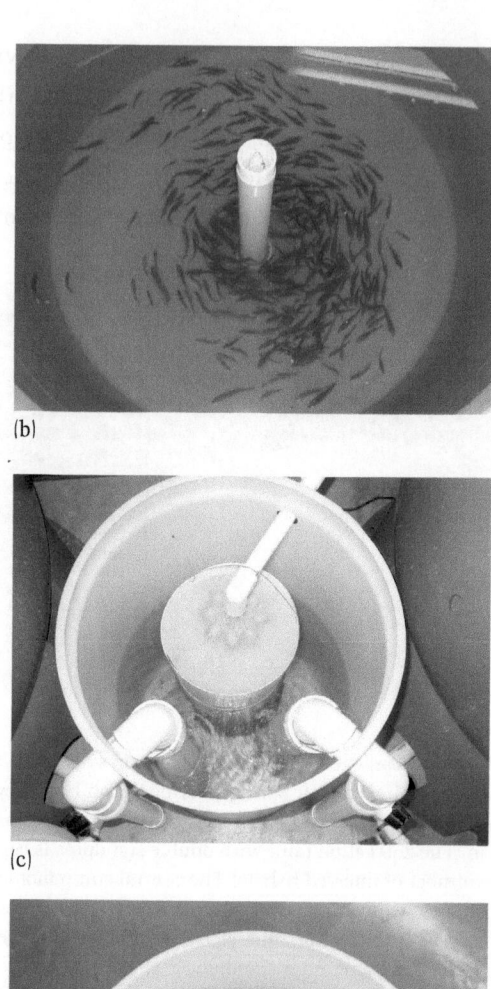

(b)

(c)

(d)

FIGURE 5.17 (*cont.*)

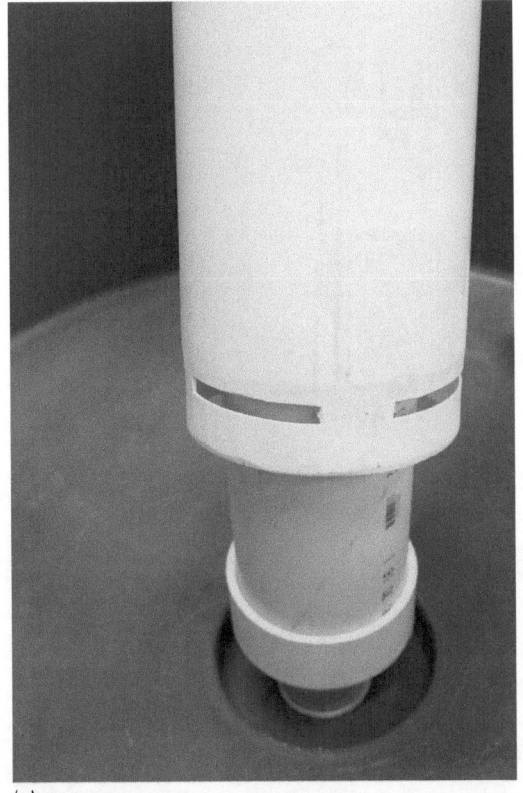

(e)

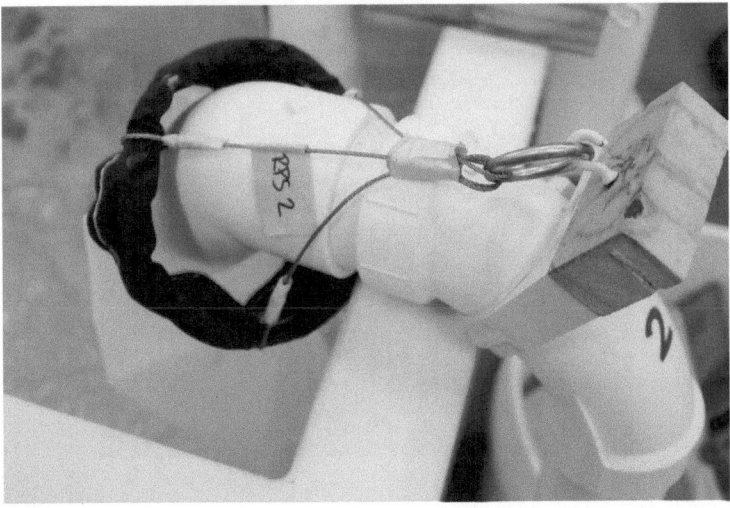

(f)

FIGURE 5.17 (cont.)

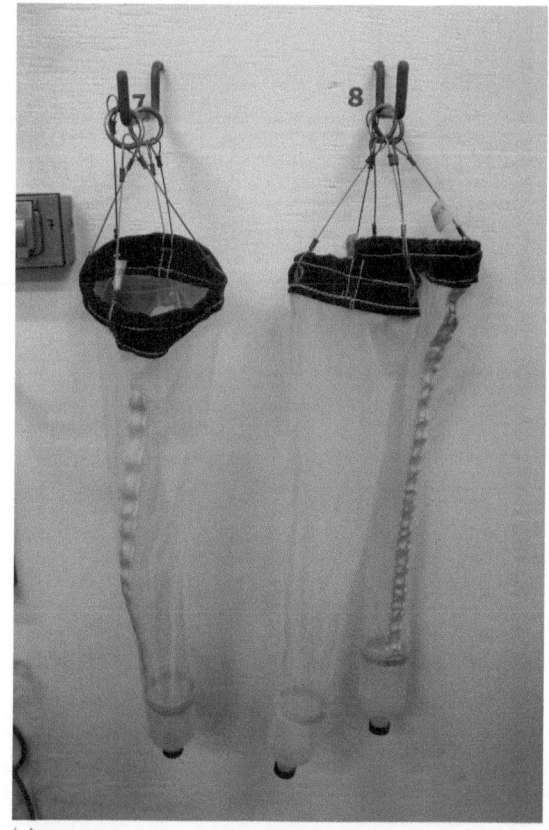

(g)

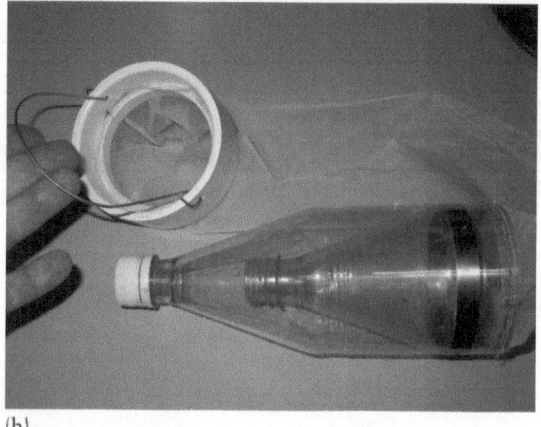

(h)

FIGURE 5.17 (*cont.*)

(Figure 5.17d, e). The water currents entrain the lightweight juvenile mussels and sweep them from the tank into a collection filter (Figure 5.17f, g, h). Benthic fishes such as darters and sculpin tend to gather on the bottom of the tank and clog the slits in the outer standpipe. To avoid an overflow, only pelagic fish species should be placed in this type of system.

Large recirculating systems like the RPS can hold larger fish species and larger numbers of fish. Consequently, they can produce large numbers of juvenile mussels. Fish species that are easily stressed in small tanks (e.g. walleye, *Sander vitreus* and freshwater drum, *Aplodinotus grunniens*) also have higher survival rates in larger tank systems. Holding all of the infested host fish in a single system is one of the major drawbacks of the large recirculating systems. A disease outbreak can spread quickly and water quality issues can kill an entire group of fish before the glochidia complete metamorphosis.

5.7.2 Flow-through Aquaculture Systems

Flow-through aquaculture systems provide an excellent environment for maintaining the health of infested host fish because waste materials are continuously being removed. Collecting newly metamorphosed juvenile mussels, however, can be a challenge. Sieves and filters designed to collect juvenile mussels tend to foul with particles in the incoming water that are similar in size to the juvenile mussels. When the filtered material is rinsed into a Petri dish for inspection and enumeration, it can be very difficult and time-consuming to separate similar-sized particles from the juvenile mussels. In some cases, there is so much material in the dish that the juvenile mussels are hard to see under the microscope and the accuracy of the juvenile count is compromised. These small particles also can foul the juvenile culture systems if they are not removed. To aid in the collection and enumeration of newly metamorphosed juveniles, it is best to hold infested hosts in a flow-through system initially and then transfer them to a recirculating system just prior to juvenile drop-off. If hosts from outside of the propagation facility's drainage area will be held in flow-through systems, it is critical to have proper filtration and effluent treatment to prevent fish escapement or the spread of fish pathogens.

5.8 *IN VITRO* METAMORPHOSIS

In vitro techniques allow for metamorphosis of the larval mussel to the juvenile stage without the aid of a live host. Instead, metamorphosis occurs in a liquid culture media consisting of salts and nutrients, fish or mammalian blood serum, lipid supplements, and a mixture of antibiotics and antimycotics (Isom and Hudson, 1982; Owen *et al.*, 2010; Lima *et al.*, 2012). Glochidia are harvested from the gravid female either non-lethally using the syringe method (Section 5.1.1) or lethally through complete excision of the brooding demibranchs (Figure 5.18a, b). Complete excision is a lethal harvest method and should only be used for species that are unimperiled. To minimize contamination of the liquid culture media, the inside of the gravid female is washed several times with sterile water and the glochidia are harvested using sterile water or sterile media. The risk of contamination can be further reduced by thoroughly cleaning and rinsing the glochidia with sterile medium and transferring a small volume of the rinse solution to the culture media.

The chances of fungal contamination increase for mussel species that take longer (>6 days) to complete metamorphosis (Owen *et al.*, 2010). Holding gravid females in captivity for long periods of time also can increase the chances of fungal contamination (Christopher Owen, personal communication). Consequently, glochidia should be placed in *in vitro* culture as soon as possible after the gravid females are collected from the wild. The culture medium also should contain a small concentration of antimycotic to control fungus. Isom and Hudson (1982) recommended the use of the antimycotic amphotericin B at a concentration of 5 µg/mL,

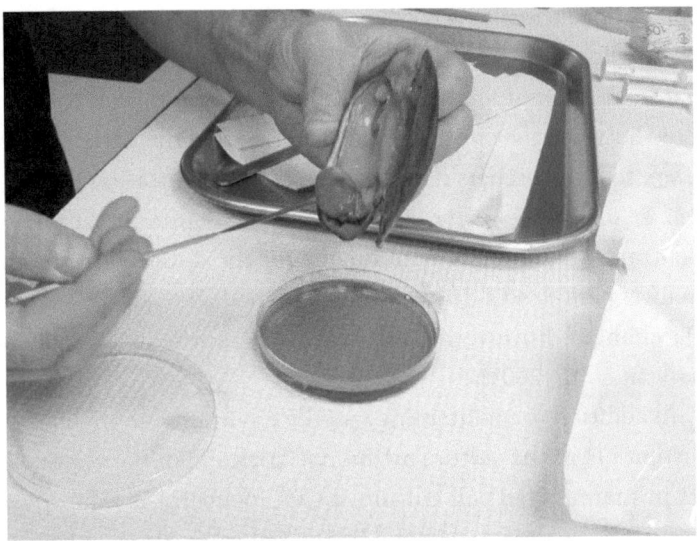

(a)

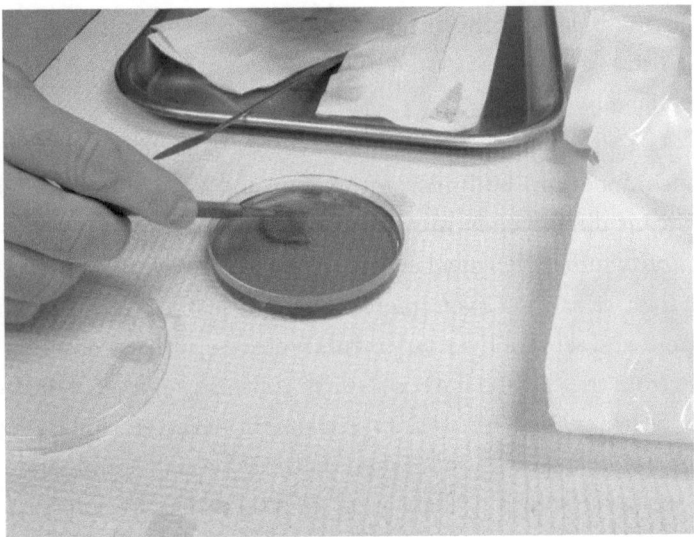

(b)

FIGURE 5.18 (a) Excision of the demibranchs for *in vitro* culture at the Warm Springs National Fish Hatchery (Warm Springs, Georgia). This lethal extraction method should not be used for imperiled species. (b) Excised gills being placed in the culture media to complete metamorphosis.
Photos: Jaclyn Zelko, USFWS. (A black-and-white version of this figure will appear in some formats. For the color version, please refer to the plate section.)

but this concentration has been found to be toxic to the rainbow mussel (*Villosa iris*), painted creekshell (*Villosa taeniata*), and elktoe (*Alasmidonta marginata*) (Owen, 2009). Amphotericin B has a small window between the effective concentration and the toxic concentration, so most researchers have reduced the concentration to 1 µg/mL (Owen *et al.*, 2010; Jaclyn Zelko, personal communication). This concentration will not control fungus for the duration of the metamorphosis period, so a daily exchange of media is required. Frequent media changes, however, may cause a stress response in glochidia (Uthaiwan *et al.*, 2001).

In addition to maintaining a sterile environment, maintaining a consistent pH in the culture medium is critical. For the maintenance of pH in mammalian cell culture, a CO_2 incubator is typically used to create a 5% CO_2 atmosphere. Pre-mixed gases also can be used to control CO_2 levels in the culture media (Roberts and Barnhart, 1999). Incubation in lower concentrations of CO_2 or even in air is possible if the culture medium is titrated to the correct pH (Kern and Barnhart, unpublished).

The source of blood serum and the associated amino-acid profile also appear to be important. Keller and Zam (1990) found no significant difference in metamorphosis success of paper pondshell (*Anodonta imbecillis*) incubated in neonatal calf serum, horse serum, trout liver, or salmon liver. Interestingly, the juveniles collected from the trout and salmon liver treatments were less active than those collected from the calf and horse serum. In contrast, survival of the freshwater pearl mussel (*Hyriopsis myersiana*) from the glochidia stage to the juvenile stage was 84% when cultured on fish serum and only 46% on horse serum (Uthaiwan *et al.*, 2001). The freshwater pearl mussel (*Hyriopsis myersiana*) also showed 94% survival on common carp serum and only 32% survival on striped catfish serum. So it seems that some mussel species may be specialists when it comes to the source of blood and others may be generalists. Regardless, each new species will likely require some experimentation to determine the ideal *in vitro* culture technique.

The advantages of *in vitro* metamorphosis include potentially higher yields of juvenile mussels, reduced overall costs, reduced labor associated with holding host fish in captivity, reduced impacts to wild fish populations, and the ability to culture freshwater mussels, even if the host is unknown (Owen *et al.*, 2010). Additionally, artificial selection for genotypes that may be adapted to particular host species may be reduced (Chris Barnhart, MSU, personal communication). The major disadvantage is that specialized equipment and training are required to prevent both microbial and fungal contamination in the culture plates. Proper aseptic technique is critical anytime the glochidia and cultures are handled in the laboratory. Key equipment needed to maintain sterile cultures include a sterile hood (ideally housed in a clean lab), a low temperature incubator, and an autoclave to sterilize equipment and glassware.

Once all the proper equipment is purchased and all the techniques are worked out, *in vitro* metamorphosis can greatly increase the yield of juvenile mussels compared to *in vivo* metamorphosis (Christopher Owen, personal communication). For the freshwater pearl mussel (*Hyriopsis myersiana*) approximately 16% of the glochidia used in an *in vivo* infestation of the common carp (*Cyprinus caprio*) were later recovered as newly metamorphosed juveniles (Uthaiwan *et al.*, 2003). By comparison, 93% of the glochidia used in *in vitro* cultures with common carp blood serum were later collected as newly metamorphosed juveniles (Uthaiwan *et al.*, 2002).

In vitro metamorphosis has been successfully completed for approximately 60 species of freshwater mussels, including 20 species listed under the Endangered Species Act of 1973 (Christopher Owen, personal communication). The technique works well with glochidia that complete metamorphosis in a short period of time, but so far has been unsuccessful for glochidia that grow during the parasitic stage (i.e. *Truncilla*, *Leptodea*, *Potamilus*, or Margaritiferidae). Further research is needed to perfect the *in vitro* technique and long-term comparisons of the health of mussels propagated *in vitro* and *in vivo* are needed.

5.9 RECOVERY OF NEWLY METAMORPHOSED JUVENILES

The final step in the process is recovery of the newly metamorphosed juvenile mussels. In some cases, it may be advantageous to extend the encapsulation period and delay juvenile recovery. For infested hosts that will be used for a free release, extending the encapsulation period can allow fish to be held in captivity until spring flood waters subside. Decreasing water temperatures in the fish holding system is the most effective method for extending the encapsulation period (Watters and O'Dee, 1999; Steingraber *et al.*, 2007). Cumulative temperature units of development (CTUD) have been used to predict the duration of encapsulation (Steingraber *et al.*, 2007). For glochidia that overwinter on the host, CTUD also could be used to predict the number of days the host will have to be held in captivity. The CTUD may not be standard across all species of freshwater mussels, so more research may be needed before applying this technique.

During *in vivo* metamorphosis, juvenile mussels must be recovered from the host holding system using either a self-cleaning tank or manual siphoning (Figure 5.19). Both methods will effectively recover juvenile mussels. For manual siphoning, a standard aquarium siphon and a fine mesh sieve are needed. Care must be taken to ensure that all of the water exiting the tank passes through the sieve. For small host species, a coarse filter on the front end may be necessary to prevent fish from getting sucked into the siphon hose. Separate siphons and sieves should be used for each tank if multiple species are being propagated. Juveniles collected on the sieve or tank filter should be rinsed into a Petri dish for enumeration.

Regardless of the juvenile recovery method used, it is critical to keep the holding system as clean as possible to facilitate juvenile counting. Excess feed, fish scales, and fish waste in the Petri dish can cause counting errors and hours of extra work. The size of sieve used to collect the juvenile mussels also will affect sample cleanliness. Use the largest mesh size that will catch the juveniles, while allowing smaller waste particles to pass. Juveniles of most mussel species can be captured on a 150 micron sieve. Some species require smaller

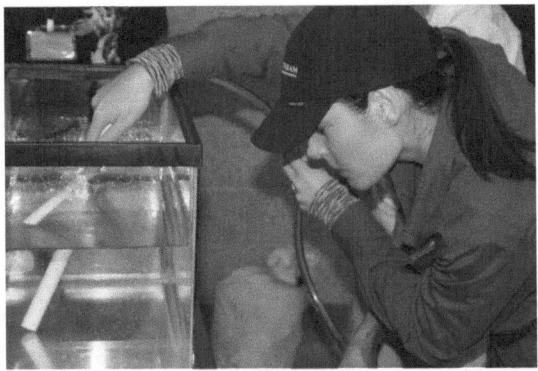

FIGURE 5.19 Manual siphoning of juveniles from static aquaria.
Photo: Matthew Patterson, USFWS.

mesh sizes and others can be captured on 200 micron or even 300
micron mesh. A series of nested sieves of different size mesh can help
remove large waste particles and provide a clean sample for counting (Gatenby, 1994; Gatenby *et al.*, 1996). For example, a 300 micron
sieve on top will catch large debris while the juveniles pass through
to the 150 micron sieve below.

During *in vitro* metamorphosis, the stages of mussel development in the culture plates can be monitored microscopically
(Figure 5.20a, b). Once metamorphosis is complete, juvenile mussels are transferred from the culture medium to fresh water. Some
researchers carry out a 25% dilution of the culture media the day
before metamorphosis is complete and then transfer the juveniles
to fresh water. Other researchers simply transfer juvenile mussels
directly from the media to fresh water. The 25% dilution does not
seem to affect the rate of metamorphosis, but the juveniles appear
more active than juveniles transferred directly to fresh water (Jaclyn
Zelko, personal communication).

5.10   ENUMERATING JUVENILES

Newly metamorphosed juvenile mussels can be enumerated by either
counting all of the juveniles or counting a subsample to estimate

(a)

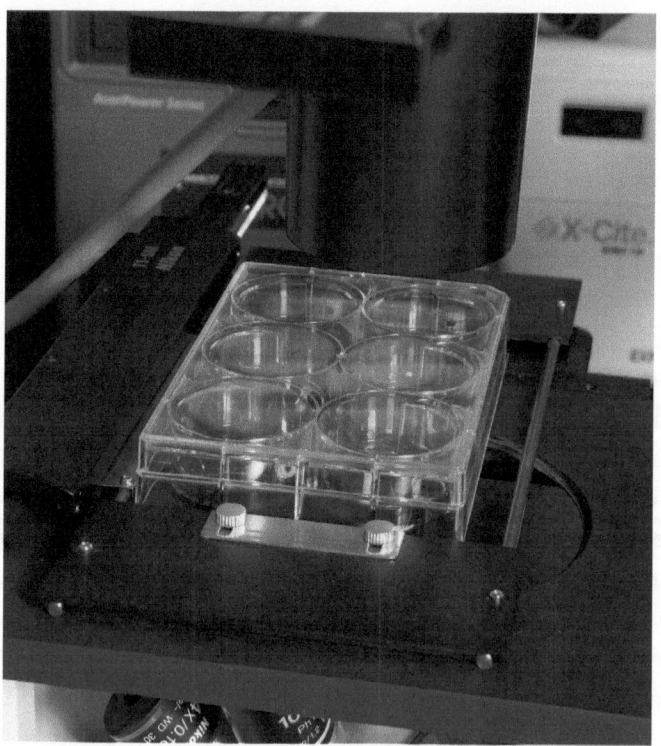

(b)

FIGURE 5.20 (a) Culture media being added to the culture plates for *in vitro* metamorphosis. (b) The culture plates can be placed under an inverted microscope to observe the stage of metamorphosis.
Photos: Ryan Hagerty, USFWS.

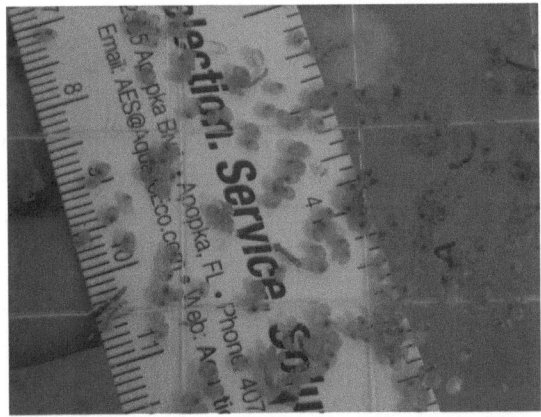

FIGURE 5.21 Juvenile mussels in a gridline Petri dish. The grid assists with the enumeration process.
Photo: Rachel Mair, USFWS. (A black-and-white version of this figure will appear in some formats. For the color version, please refer to the plate section.)

the total number. The number of juveniles recovered, the species of mussel, and the future use of the juveniles will help determine the best enumeration method. A complete count of all juvenile mussels can be very time-consuming, especially if a large number of juveniles is recovered. A complete count, however, may be required for imperiled species or juvenile mussels that will be used for research. If a complete count is necessary, place the juveniles in a Petri dish with gridlines that divide the dish into multiple quadrants (Figure 5.21). Systematically moving from one quadrant to another will help minimize counting errors. Without the gridlines, individual mussels may be counted more than once or entire areas of the dish may be overlooked. Counting errors also can be minimized by keeping track of the number of juveniles on a hand counter. The amount of water in the Petri dish also can affect the accuracy of juvenile counts. Too little water can cause wave action in the dish, washing juveniles across the gridlines. To prevent wave action, fill the Perti dish about one-half to three-quarters full with water.

If an exact count is not necessary, the number of recovered juveniles can be estimated using the volumetric method for enumerating glochidia described in Section 5.5. Estimates are best used for large numbers of juveniles when a complete count is not feasible. Estimates become less reliable as the number of juveniles decreases. A complete count is more precise when the number of juveniles is less than 5000.

6 Juvenile Mussel Culture

Rachel A. Mair

Significant improvements have been made in the field of freshwater mussel propagation since 2010. Mussel culture has progressed from releasing hundreds of thousands of small, two-month-old juvenile mussels (less than 5 mm in length) to releasing thousands of large, tagged, subadult mussels (Figure 6.1). With that being said, new methods are being developed on an annual basis and culture systems in use one year may be replaced the following year. Every propagation facility is different, so a culture system that works well at one facility may not work

FIGURE 6.1 Subadult mucket (*Actinonaias ligamentina*) tagged for release.
Photo: Matthew Patterson, USFWS. (A black-and-white version of this figure will appear in some formats. For the color version, please refer to the plate section.)

well at another. Each mussel species also will have different culture requirements (i.e. food, flow, substrate, etc.), so a culture system that works well for one species may not work for another. The objective of this chapter is to provide a review of all the freshwater mussel culture techniques currently in use. It will be up to each facility to decide which culture system works best for them and their target species.

6.1 INDOOR CULTURE SYSTEMS

Facilities that culture freshwater mussels indoors typically either do not have access to ponds or other "wild water" sources or culture juvenile mussels initially in indoor recirculating systems before moving them to outdoor culture systems for grow-out. Without access to a wild water source, indoor culture requires either the production of algae on-site or the purchase of commercially available algae. The benefits of indoor culture include the ability to isolate different mussel species or isolate mussels of the same species from different river drainages. Isolation can help reduce the risk of spreading disease and/ or non-native species (see Table 6.1 for a list of advantages and disadvantages of indoor culture).

The following section will provide a description of the main types of indoor mussel culture systems currently in use; including Barnhart buckets (modified downwelling culture systems that will be called bucket systems from this point forward), rearing pans, static culture chambers, partial flow-through bucket systems, and upwellers. This list is not exhaustive because new systems are being created or adapted on an annual basis. As mentioned above, every facility is different and a system that might work for one facility may not work for another. Each facility might have different mussel species, water quality, water quantity, food availability, personnel, etc. Each mussel species is likely to have its own physiological and nutritional requirements, but many of these species-level differences are either unknown or unreported in the literature. Determining the best culture system for any given facility can sometimes be trial and error and it may take several propagation seasons to get good and consistent juvenile growth and survival.

Table 6.1. *Advantages and disadvantages of indoor culture of freshwater mussels*

Advantages of indoor culture	Disadvantages of indoor culture
Keep species separate	Unless wild water is available, must purchase or culture algae as the primary food source
Keep river drainages separate Prevents the spread of organisms from one system to another	Mussels may need more than just algae to grow and survive well. Outdoor water sources typically have numerous species of algae, zooplankton, and bacteria Observed growth rates are often significantly higher in outdoor culture
Easily monitor health and survival by visual inspection Control of the culture environment (temperature, amount and type of food, etc.)	Very labor intensive Requires increased electricity to run pumps and chillers year round. More funding necessary
Low risk of a pollution event	Requires a lot of space Reliant on electricity; pump failure can cause mortality when staff are away from the lab

6.2 PRIMARY INDOOR CULTURE SYSTEMS FOR NEWLY METAMORPHOSED AND YOUNG JUVENILES

Not all indoor culture systems are suitable for culturing newly metamorphosed or very young juvenile mussels. In the early stages, juvenile mussels are extremely fragile and easily stressed by handling. They are vulnerable to small invertebrate predators (i.e. flatworms, ciliates, dipterans, etc.) that can enter the culture system through the water source, host fish, food source, or culture sediment. The small size of young juvenile mussels also makes them very lightweight and easily washed out of the culture system. Consequently, growth and survival can be very tentative and escapement high at this stage.

Over-feeding at this early stage can also lead to decreased growth and survival. Excess feed can foul the culture system, creating water quality problems (i.e. ammonia) and making it more difficult for juvenile mussels to access food. The feeding apparatus has not fully developed, making it difficult for them to access and obtain enough nutrition for growth and survival (Gatenby, 2000). Many propagation facilities have observed significant improvements in both growth and survival in recent years mostly due to improvements in delivering suitable diets on a continuous basis (24 hours a day) and utilizing culture systems that support juvenile mussel feeding strategies (Mair, 2013). Once juvenile mussels reach approximately 4 months of age or over 1 mm in length, mortality usually decreases.

It is critical to design culture systems that take into account the vulnerabilities of this early life stage. Some of the culture systems currently in use for newly metamorphosed and young juveniles include the bucket system, rearing pan, static chamber, partial flow-through bucket systems, and beaker/pulse flow system.

6.2.1 Bucket System

The bucket system, developed by Dr. Chris Barnhart at Missouri State University (Springfield, Missouri), is a modified downweller modeled after the marine shellfish industry (Figure 6.2). This culture system works very well for culturing juvenile mussels from 1-day-old (Barnhart, 2006) up to about 5 mm in length or until growth rates start to decline, whichever comes first. Each system consists of two buckets, a 18.9 L lower bucket with a 11.4 L upper bucket nested inside (Barnhart 2006; Mair 2013; Figure 6.3a). The upper bucket has a series (typically 5 to 7) of 1.5 inch (38 mm) or 2.0 inch (51 mm) holes drilled in the bottom. The PVC culture chambers that house the juvenile mussels fit tightly inside the holes in the upper bucket (Figure 6.3b). The culture chambers are made from nylon mesh glued onto a small section of 1.5 inch or 2.0 inch schedule 40 polyvinyl

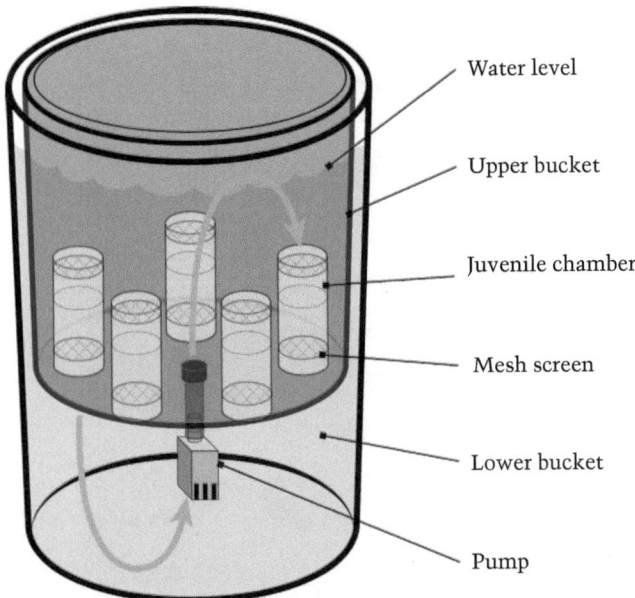

FIGURE 6.2 Diagram of the bucket system showing the upper and lower buckets, juvenile chambers with mesh screening, submersible pump, and the direction of water flow.
Diagram adapted from Mair (2013) by Kristin Simanek, USFWS.

chloride (PVC) pipe. The pipe and mesh are pressed into the appropriate sized PVC coupling, forming a cup (Figure 6.3c, d, e). The two cups are pressed together to create a chamber with mesh on the top and bottom, preventing juvenile escapement. A total of 18 L of water is recirculated between the two buckets. Water is pumped from the lower bucket into the upper bucket, using a small submersible mini-jet 404 pump (Figure 6.3f; Marineland, Inc., Blacksburg, Virginia) and downwells through the culture chambers into the lower bucket.

For newly metamorphosed juveniles, 150 micron nylon mesh is typically used in the culture chambers. It is important that all culture chambers in a single bucket system have the same mesh size. Different mesh sizes in the same system will lead to different flow rates in each chamber (the chambers with finer mesh experiencing decreased flows). The fine mesh screens will foul over time, decreasing water

(a)

FIGURE 6.3 (a) The bucket culture system designed by Dr. Chris
Barnhart (Missouri State University). The upper bucket with attached
submersible pump and the lower bucket pulled apart for demonstration.
(b) The underside of the upper bucket showing the insertion of the
mesh culture chambers. (c) A mesh culture chamber that holds the
juvenile mussels. (d) The two sections of the mesh culture chamber
separated for demonstration. (e) The inside of the mesh culture
chambers showing the micron mesh. (f) The underside of the upper
bucket showing the submersible pump attachment.
Photos: Ryan Hagerty, USFWS.

(b)

(c)

FIGURE 6.3 (*cont.*)

(d)

(e)

FIGURE 6.3 (*cont.*)

(f)

FIGURE 6.3 (*cont.*)

flow inside the culture chamber. To maximize flow, the culture chambers should be sprayed forcefully with a pressurized garden sprayer
(see Chapter 3; Figure 3.4) or hose with spray nozzle (Figure 6.4)
1–2 times per day. The entire bucket system also should be disassembled, thoroughly cleaned, and replenished with fresh water once per
week. Growth and survival of the juvenile mussels can be assessed
during the weekly cleaning process. As the mesh screens age, they can
become difficult to clean. If this occurs, the screen should be replaced
or carefully bleached and rinsed. If bleach is used, the chlorine should
be neutralized with sodium thiosulfate and then rinsed thoroughly

FIGURE 6.4 A garden hose nozzle sprayer with an adjustable spray pattern to clean the mesh sieves in the juvenile culture chambers. Photo: Rachel Mair, USFWS.

before juveniles are added. As the juveniles grow, it is very important to increase the screen size on the containers to increase flow rate, improve food delivery, and minimize clogging. Once the mesh size reaches between 400 and 1000 microns, and/or juvenile growth rates begin to decline, mussels should be moved to a secondary culture system such as outdoor baskets, outdoor cages, rearing pans, or upwellers. Growth rates typically start to decline in the bucket systems when juveniles reach approximately 2–5 mm in length.

6.2.2   Rearing Pan System

The rearing pan system was developed at the Virginia Department of Game and Inland Fisheries' Aquatic Wildlife Conservation Center (AWCC) in Marion, Virginia. This system was originally designed for rearing newly metamorphosed juvenile mussels, but the system also can be used for older juveniles. The system consists of a series of 5-quart plastic culture pans (Dare Products, Battle Creek, MI) with a

0.5 inch PVC bulkhead and 0.5 inch PVC standpipe (Figure 6.5a, b). Water exiting the standpipe drains through a 150 micron mesh bag into a reservoir sump located below the pan (Figure 6.5c). The mesh bag is designed to catch any newly metamorphosed juveniles that might escape. As the juveniles get larger, fewer escapements occur. Water enters the pan at an angle, creating a circular flow of about 1.0–1.5 L per minute. Each pan is filled with 200 mL of fine white play sand less than one millimeter in diameter (Figure 6.5d). The rearing pans can be run as a flow-through system with pond water or as a recirculating system. If the system is recirculating, 50% water changes should be performed once or twice a week, depending on the feeding rate. The system can be cleaned periodically by swirling the pans and pouring off any accumulated waste and excess feed. If the juveniles are small, pour the water through a sieve to catch any mussels suspended in the water column during the swirling process. A 200 micron sieve works well for 1-week-old juveniles. If the system is running flow-through pond water, cartridge pre-filters (30–100 micron) should be used with newly metamorphosed or very young juveniles. The pre-cartridge filters will help prevent potential predators from entering the culture system. Fine mesh filters will clog and need to be changed regularly. Once the juveniles reach 1 mm in length, a larger mesh filter can be used and changed or cleaned as needed.

6.2.3 Static Culture Chambers

Static culture chambers also have shown good growth and survival for newly metamorphosed juvenile mussels. A thin layer of fine sediment (less than 2 mm) is placed on the bottom of an aquarium, plastic container, or bucket with an air stone (Figure 6.6a, b, c). Fine sediment can be collected along river banks or near large springs. Fine sediments should not have an offensive odor and should not need to be autoclaved or sterilized. Some fine sediment, however, may have flatworms, flatworm eggs, and other organisms that could prey on juvenile mussels. If you are unsure about the sediment source, the best practice is to autoclave or microwave sediments to kill unwanted

(a)

(b)

FIGURE 6.5 (a) Wide angle view of the rearing pan system at the
Virginia Department of Game and Inland Fisheries' Aquatic Wildlife
Conservation Center (Marion, Virginia). (b) A close-up of one of the
rearing pans showing the standpipe and circular water flow. (c) The
rearing pan system showing the rearing pans, substrate, and mesh bags
for collecting juveniles that may escape from the system. (d) A single
rearing pan with juvenile freshwater mussels feeding at the surface.
Photos: (a, c) Nathan Eckert, USFWS. (b) Matthew Patterson, USFWS.
(d) Rachel Mair, USFWS. (A black-and-white version of part (d) of this
figure will appear in some formats. For the color version, please refer to
the plate section.)

(c)

(d)

FIGURE 6.5 (*cont.*)

(a)

(b)

FIGURE 6.6 (a) Hruska boxes stacked in a laboratory incubator. (b) Hruska boxes at Dr. Chris Barnhart's lab (Missouri State University). (c) Juvenile mussels making trails in the substrate in a Hruska box. (d) Artemia boxes can be used to clean and separate mussels from sediment substrate in the Hruska box culture chambers. (e) Water and sediments in the Hruska box can be passed through a sieve during the weekly cleaning process.

Photos: (a, d) Frankie Thielen, Fondation Hëllef fir d'Natur, Luxembourg. (b) Chris Barnhart, Missouri State University. (c, e) Beth Glidewell, Missouri State University. (A black-and-white version of part (c) of this figure will appear in some formats. For the color version, please refer to the plate section.)

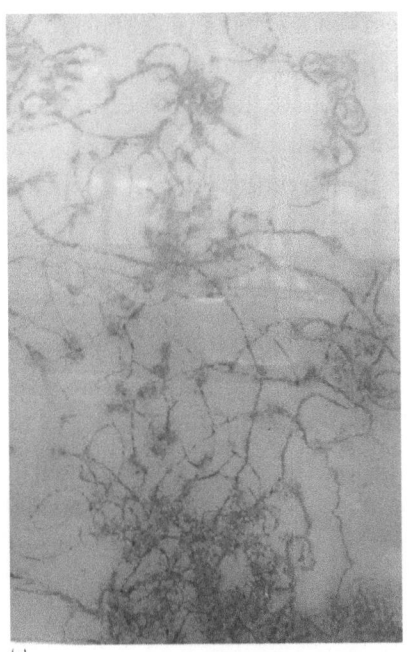

(c)

(e)

(d)

FIGURE 6.6 (*cont.*)

organisms (Gatenby, 1994). Sediments should be sieved to ensure the particles are not the same size as the juvenile mussels. Pre-sieving the sediment will make it easier to separate juvenile mussels from the sediment particles, improving the accuracy of sample counts and growth rate estimates. For newly metamorphosed juveniles, only use sediment particles that pass through a 200 micron sieve. Water and sediment should be changed weekly. When the juveniles reach 500 microns in length, water exchanges can occur every 2 weeks or more.

One example of a static culture chamber is the Hruska box (Hruska, 1992). These boxes are widely used in Europe to culture juveniles in the family Margaritiferidae. Juveniles in this family can be very fragile and do not seem to fare well in the turbulence of the downwelling bucket systems. The boxes are filled with fine sediment, 500 mL of static water, and approximately 250 newly metamorphosed juveniles. The boxes are cleaned weekly by passing the water and sediment through a sieve with 180 micron mesh (Figure 6.6d, e). Juvenile mussels collected on the sieve are then transferred to a new box with fresh water and sediment. Juveniles are fed a diet of Shellfish Diet and *Nannochloropsis* (Reed Mariculture, Inc.) mixed with 10 L of river water. At the Mill of Kalborn, Luxembourg, Europe, juvenile mussels in the boxes are fed a single concentration of algae the first month (120 microliters of Shellfish Diet and 4 drops of *Nannochloropsis*), a double concentration of algae the second month (240 microliters of Shellfish Diet and 8 drops of *Nannochloropsis*), and a triple concentration the third month (360 microliters of Shellfish Diet and 12 drops of *Nannochloropsis*) until mussels reach 1 mm in length (approximately 100 days) (Frankie Thielen personal communication; Eybe *et al.*, 2013).

6.2.4   Partial Flow-through Bucket System

Developed by Bryan Simmons (USFWS) while working for the Kansas Department of Wildlife and Parks, the partial flow-through bucket system is a combination of the rearing pan framework and the bucket system that uses pond water as a food source (Figure 6.7a, b). Water is pumped from the pond into a 125 gallon conical bottom head tank

(a)

FIGURE 6.7 (a) The original partial flow-through bucket system, designed by Bryan Simmons while working for the Kansas Department of Parks and Wildlife, combined the rearing pan framework with the bucket system. (b) A view inside the buckets of the partial flow-through system. The system was designed to use pond water as a food source instead of an artificial diet. (c) Partial flow-through bucket system at the Alabama Department of Conservation and Natural Resources' Alabama Aquatic Biodiversity Center (Marion, Alabama). (d) The inside of the partial flow-through bucket system showing the juvenile culture chambers. (e) Juvenile Alabama lampmussel (*Lampsilis virescens*) inside the partial flow-through bucket system culture chamber.

Photos: (a, b) Bryan Simmons, USFWS. (c, d, e) Paul Johnson, Alabama Aquatic Biodiversity Center. (A black-and-white version of part (e) of this figure will appear in some formats. For the color version, please refer to the plate section.)

(b)

(c)

FIGURE 6.7 (*cont.*)

(d)

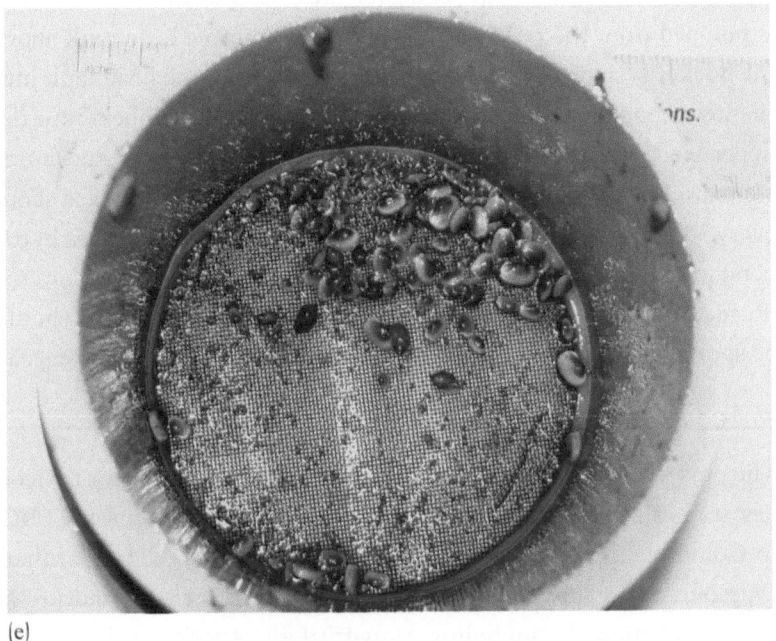

(e)

FIGURE 6.7 (*cont.*)

through a pair of canister filters. Filtration of the pond water removes potential invertebrate predators that may be present in the pond water. The pump is connected to a timer that refills the head tank every 30 minutes (the refill interval can be adjusted depending on the number of buckets and the flow rate to each bucket). Periodic filling allows high sediment loads to settle to the bottom of the head tank.

Water gravity-feeds from the head tank to the individual bucket systems through a PVC manifold. The individual bucket systems are constructed as described in Section 6.2.1 with one modification. The bottom bucket is outfitted with a bulkhead drain to allow for flow-through. The drain must be set at a sufficient height to ensure the water level in the upper bucket covers the culture chambers. The water exiting the drain enters a gutter system and returns to the pond.

The Alabama Department of Conservation and Natural Resources' Alabama Aquatic Biodiversity Center (Marion, Alabama) modified this system for upwelling (Figure 6.7c, d, e). In this case, water is pumped from the pond to a sump and then into a head tank above the bucket systems. Water then gravity-flows from the head tank into the lower bucket and upwells through the culture chambers into the upper bucket. A standpipe in the center of the upper bucket drains water back to the sump. All the water in the system is exchanged about every 30 minutes with fresh pond and well water. Food particles in the pond water are supplemented with Shellfish Diet and *Nannochloropsis* (Reed Mariculture, Inc.). Similar to the bucket systems, screens should be sprayed often and exchanged or bleached and rinsed when clogged.

6.2.5   Pulsed Flow-through System

The pulsed flow-through system is based on systems used for toxicology studies at the Columbia Environmental Research Center (CERC) in Columbia, Missouri. James Kunz (CERC) and Dr. Chris Barnhart (Missouri State University) adapted the systems for early culture of freshwater mussels, including *Margaritifera*, *Anodonta*, *Lampsilis*, *Quadrula*, and others (Barnhart, 2015). Juvenile mussels are held in glass or plastic culture beakers (250–2000 mL) with a fine layer of sand to provide habitat (Figure 6.8a). A pulse of flow is delivered into the beakers every 60–90 minutes to exchange water, remove waste, and introduce food. The system is best suited to relatively small batches of juveniles (e.g. 1000 newly metamorphosed juveniles per 250 mL beaker).

The system used at Missouri State University includes two buckets, one to mix water and one to deliver food (Figure 6.8b). The

(a)

(b)

FIGURE 6.8 (a) The glass culture beakers and sand substrate in the pulsed flow-through system at Missouri State University. (b) The pulsed flow-through system at Missouri State University. The head tank and mixing reservoir are shown on the top shelf with the peristaltic pump and culture beakers below. (c) Close-up of the culture beakers and water/feed delivery manifold in the pulsed flow-through system. (d) The nozzle and narrow tubing mounted in silicone stoppers used to deliver water and feed to the culture beakers.

Photos: Chris Barnhart, Missouri State University.

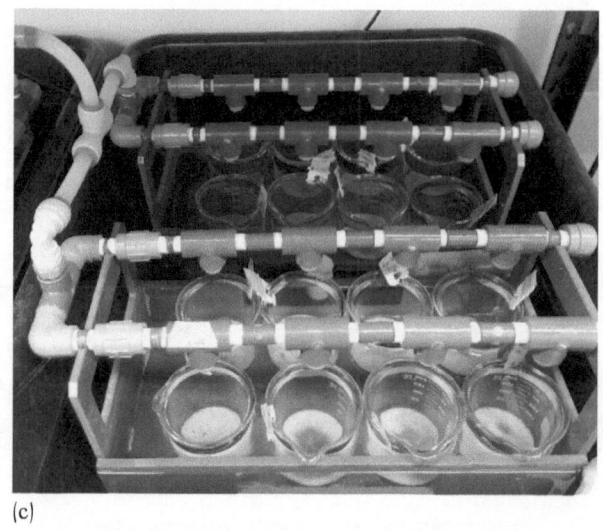

(c)

(d)

FIGURE 6.8 (*cont.*)

first bucket (head tank) continuously receives water from a source and is outfitted with a standpipe to maintain the appropriate water level. The second bucket (mixing reservoir) receives water from the head tank as well as food from a peristaltic pump. The mixing reservoir refills and delivers water periodically to an array of culture beakers via a solenoid valve, which is controlled by a timer. During each feeding event enough water is released from the mixing reservoir to exchange the full volume of each beaker. The beakers overflow and the waste water is channeled to a discharge point. The typical feeding interval is one exchange per hour. The interval between feeding events can be adjusted as long as the rate of food delivery and the refill water flow between the head tank and mixing reservoir are adjusted appropriately to deliver the proper food concentration.

The manifold that distributes water and food to the culture beakers is constructed of 0.25 inch schedule 80 PVC pipe and fittings (Figure 6.8c). The water is delivered through "nozzles" made of short (3 cm) lengths of stainless hypodermic tubing (18 gauge) mounted in silicone stoppers (Figure 6.8d). Suitable stoppers are Fisher silicone #097041D (10 × 15 × 25 mm). The narrow nozzles provide more resistance to flow than the path through the manifold, so each beaker receives a similar amount of water. The narrow nozzles also reduce the volume of flow, but increase the speed, so that the water is delivered to the bottom of each beaker without creating excessive turbulence and stirring up the sediment. A potential drawback of the narrow opening is that any debris in the delivery water can clog the nozzles. Therefore, the nozzles are removed and cleaned weekly, and the entire system is disassembled and cleaned if frequent clogging becomes an issue. It is advisable to run each manifold through at least one feeding cycle with empty beakers to ensure each beaker receives a similar delivery rate.

This system has proven effective for raising the early life stages of *Margaritifera* sp. and various species in the family Unionidae to a size sufficient for research or transfer to secondary culture. It also appears to produce good growth and survival for delicate species. Several advantages of this system include: individual beakers are independent units and easily removed for examination and cleaning, juvenile mussels are provided with substrate for burrowing, and there are no screens to clog. Different beaker diameters may require different manifold dimensions. As with the bucket systems, growth appears to slow when mussels reach a size suitable for secondary culture.

6.3   SECONDARY INDOOR CULTURE SYSTEMS

As juvenile mussels grow to a larger size (greater than three millimeters in length), they should be transferred to a secondary culture system. In addition to the tank and raceway upweller systems described below, rearing pans and partial flow-through buckets also can serve as secondary culture systems.

6.3.1   Tank Upweller

First developed by the marine industry for clam and oyster culture, the tank upweller has been adapted and tested for use in freshwater mussel culture (Mair, 2013). This relatively simple system is very effective for culturing juvenile mussels greater than 3 mm in length and can be set up as recirculating in the laboratory or flow-through beside or inside a pond (Mair, 2013). A tank upweller typically consists of a large, rectangular main tank, juvenile culture chambers, side drains for each chamber, and a sump (Figure 6.9a, b. c). Culture chambers can be constructed using 6–12 inch PVC or a 5 gallon bucket with the bottom removed (Figure 6.9d). A hole (1.5 inch or larger) is cut into the side of the culture chamber and nylon mesh is glued to the bottom. Water and food are pumped from the pond or sump (if recirculating) into the main tank and exit through the side drain. The side drains are connected to the culture chamber, so the only way for the water to exit the system is to upwell through the culture chambers (Figure 6.9e). The tank upweller at Missouri State University was constructed with a utility sink and a large sump (Figure 6.9f). The utility sink sits inside the sump and holes drilled in the bottom house the juvenile culture chambers (Figure 6.9g). Water and algae

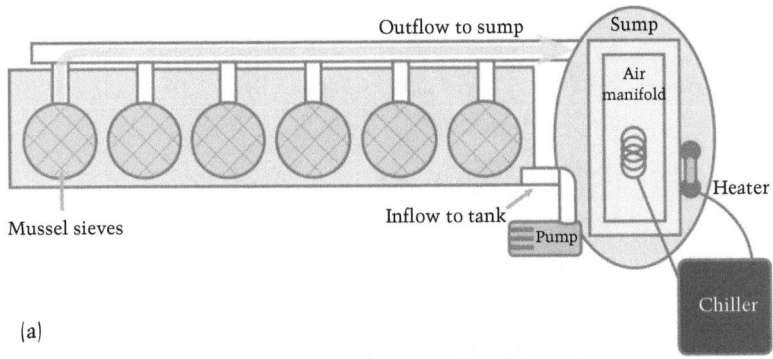

(a)

(b)

FIGURE 6.9 (a) Diagram of an indoor, recirculating tank upweller from above. Temperature-controlled water is pumped from the sump into the rectangular tank and then upwells through the mussel culture chambers. Water exits the culture chambers through the side drain and travels back to the sump. The air manifold keeps the algae in suspension. (b) The recirculating tank upweller at White Sulphur Springs National Fish Hatchery. Water and food are pumped into the main fiberglass tank, upwell through the 12 inch PVC culture chambers, exit out the side drain, and return to the sump. (c) A view from above the culture chambers showing the sub-adult mussels resting on the mesh screen. (d) Diagram of the mussel culture chamber for the tank upweller with dimensions. (e) Side view of the tank upweller. Water is pushed up into the mussel culture chambers and then out the side drains and back to the sump.

(c)

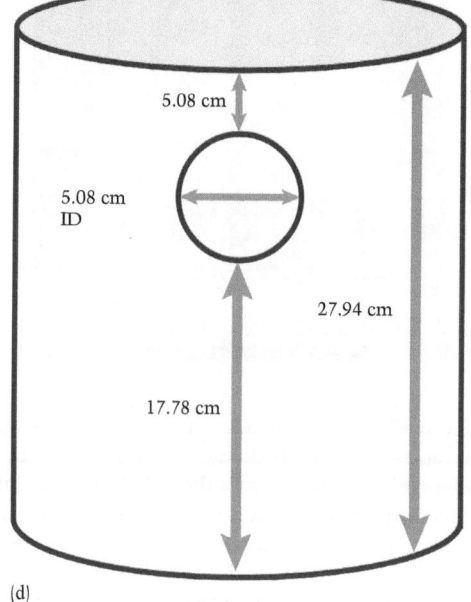

5.08 cm

5.08 cm
ID

27.94 cm

17.78 cm

(d)

FIGURE 6.9 (*cont.*) (f) The tank upweller at Missouri State University. Holes are cut in the bottom of the white utility sink to hold the juvenile culture chambers. Water is pumped into the sump, upwells through the culture chambers, and returns to the sump through a drain in the center of the utility sink. Water also can be pumped up to the pan system on the wall for additional rearing space. (g) Close-up of the juvenile culture chambers in the tank upweller at Missouri State University.
Diagrams: (a, d, e) Adapted from Mair (2013) by Kristin Simanek, USFWS. Photos: (b, c) Rachel Mair, USFWS. (f, g) Bryan Simmons, USFWS.

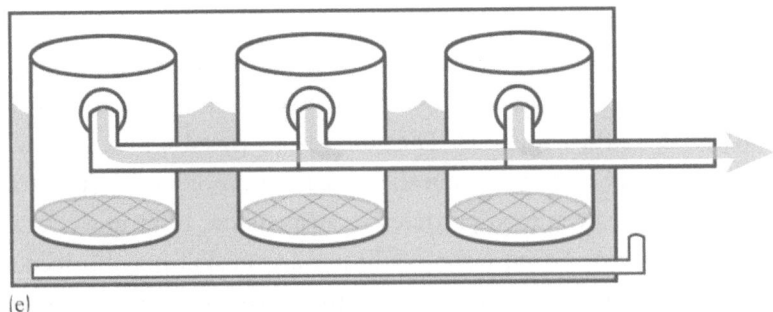
(e)

(f)

(g)

FIGURE 6.9 *(cont.)*

are pumped into the sump, upwell through the culture chambers, and return to the sump through the center drain in the utility sink.

The size of the nylon mesh in the juvenile culture chambers is typically 600–1000 microns. The sump is outfitted with a 0.5 inch PVC air manifold to maintain dissolved oxygen concentrations at saturation and keep food particles in suspension. If an upweller is set up as a recirculating system, a heater and chiller may be needed to maintain consistent water temperatures. Recirculating upwellers should receive a 50% water change 1–2 times per week and all mesh screens should be cleaned or scrubbed with a brush during the water change. A biofilter (i.e. mesh bag with bioballs or ceramic rings) also can be added to the sump to help control ammonia build-up. The flow rate inside the culture chambers is typically 2–5 gallons per minute. Mussels can be cultured in upwellers until they reach 5–30 mm in length. As long as growth rates remain consistent, mussels can remain in upwellers until tagging and release. Ideally, mussels should be moved to an outdoor culture system to increase growth rates and reduce the time to release once they reach 5–10 mm in length.

6.3.2   Raceway Upweller

The raceway upweller is a relatively simple way to convert an existing hatchery raceway into a mussel culture system using items that can be found at most hardware stores. The design is basically an elaborate drain system for the raceway. Water entering the raceway must upwell through the mussel culture chambers to exit through a 4 inch PVC drain (Figure 6.10a, b). The raceway upweller designed by Missouri State University and housed at the Kansas City Zoo is a flow-through system that draws water from a 20 acre lagoon (Figure 6.10c, d). The flow rate in the culture chambers is approximately 1 cm/s. If additional flow is needed, a larger pump can be installed or the number of exit drains can be reduced.

The juvenile culture chambers are constructed with inverted 5 gallon buckets. The bottom of the bucket is removed and a 2.0 inch drain hole is drilled in the side wall. The drain hole must be positioned

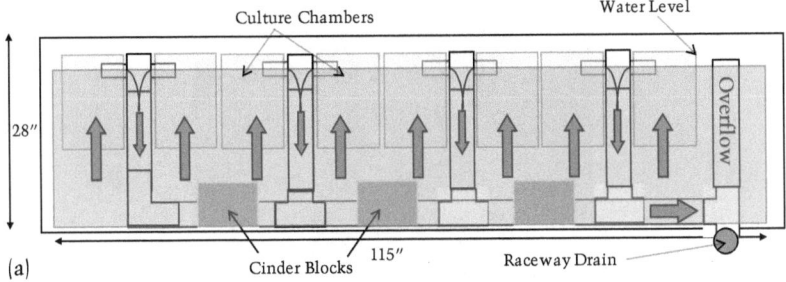

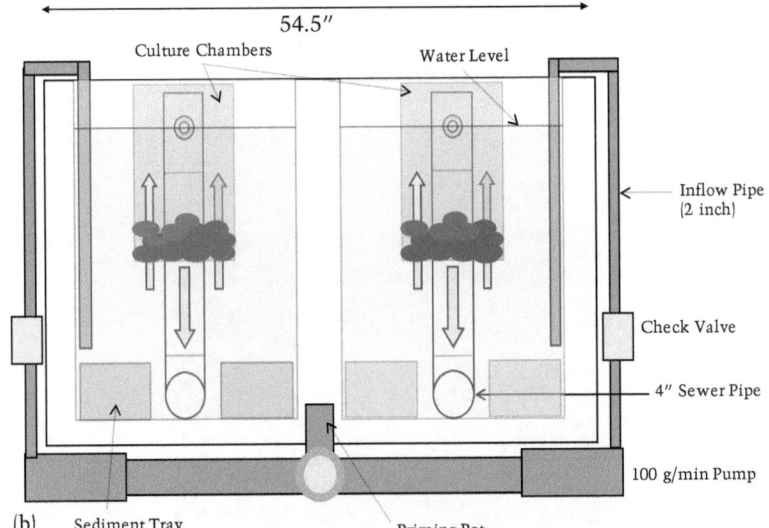

FIGURE 6.10 (a) A side view diagram of the raceway upweller system
at the Kansas City Zoo. Water is pumped into the raceway, upwells
though the culture chambers and exits through the 4 inch drain pipe.
Cinder blocks are used to keep the entire system on the bottom of
the raceway. (b) Cross-sectional view of the raceway upweller system
showing the inflow pipe. (c) A concrete hatchery raceway outfitted with
upwelling culture chambers (left) and a second raceway deconstructed
for cleaning (right). (d) A close-up view of the cinder blocks and drain
system for the raceway upweller. (e) Juvenile pink mucket (*Lampsilis
abrupta*) being reared in the raceway upweller at the Kansas City Zoo.
Diagrams: (a, b) Bryan Simmons, USFWS. Photos: (c, d) Bryan Simmons,
USFWS. (e) Chris Barnhart, Missouri State University. (A black-and-
white version of part (e) of this figure will appear in some formats. For
the color version, please refer to the plate section.)

(c)

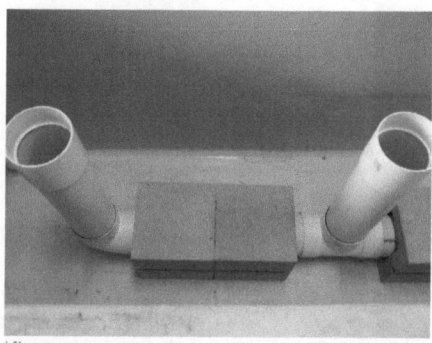

(d)

(e)

FIGURE 6.10 (*cont.*)

below the top of the raceway wall and below the expected water line to function properly. The center of the bucket lid is removed, leaving only the snap ring. The snap ring is used to hold the mesh screen on the bottom of the culture chamber. Depending on the flow rate and available food, large numbers of mussels can be placed in each culture chamber (Figure 6.10e). Each set of two culture chambers drains into one standpipe and rests on a single 2.0 inch pipe with a hole in the center. An overflow drain is added as a safety precaution to ensure the unit will not overflow if the mesh screens clog.

6.4 OUTDOOR CULTURE SYSTEMS

Many culture facilities have experienced excellent juvenile growth and survival in outdoor culture systems that use a wild-water source. Higher rates of survival have been observed when juveniles greater than 5 mm in length are deployed to outdoor culture systems. Space limitations, along with the labor-intensive nature of indoor culture systems, however, often lead to the deployment of juvenile mussels that are between 1 and 5 mm in length or smaller. Many of the indoor culture systems discussed previously can be used outdoors or can be plumbed with flow-through pond or wild water. Outdoor culture systems can be deployed in ponds, lakes, impoundments, and even low-flow areas of rivers and streams. The outdoor culture systems discussed below include floating cages, standing cages, floating baskets, silos, bunkers, floating upweller systems (FLUPSYs), and suspended upweller systems (SUPSYs).

6.4.1 Cages

One of the first cages used to culture freshwater mussels was designed by the United States Bureau of Fisheries in the early 1900s (Howard, 1922). Genoa National Fish Hatchery (Genoa, Wisconsin) modified this design and became the first facility in recent history to utilize cage culture on a large scale. The Genoa-style mussel cage consists of a 2 foot × 3 foot × 1.5 foot angle aluminum frame, 0.5 inch galvanized aluminum screen fastened to the frame with rivets, and a plywood base (Brady *et al.*, 2012; Figure 6.11a). Fine sand (approximately 1–2 inches in depth) is placed on the plywood bottom to serve as burrowing substrate for the juvenile mussels (Figure 6.11b). Cages can be used for initial juvenile culture or long-term grow-out. For initial culture, infested fish are placed directly into the cage and then released after the juvenile mussels drop off. For long-term grow-out, larger juvenile mussels are placed directly into the cage. Depending on conditions at the culture location, the cage can remain in place for 1 year or longer. Mussels are then removed from the cage, tagged, and released to a restoration site.

The cages are heavy so they are typically deployed by a team of people with the assistance of SCUBA divers. The best juvenile growth and survival occur when cages are deployed in eutrophic, well-oxygenated waters with minimal flow. Regardless of flow rate, a portion of the juvenile mussels will likely escape from the cage. When mussels are removed from the cage for tagging and release, it is important to check both underneath and downstream of the cages for potential escapees. In areas of high flow, all of the mussels could be swept from the cage and washed downstream of the cage location. In fact, the entire cage can be swept downstream in extreme high-flow events. In these high-flow areas, the preferred cage system would be the mussel silos or mussel bunkers discussed later in this chapter.

Cages can be customized for different habitats, host species, and mussel species. They can be suspended in the water column using legs or a floating rack in habitats where the bottom may experience periods of anoxia (Figure 6.11c, d). For small host fishes like minnows and darters, the cage can be wrapped with 0.25 inch screen to prevent escapement. For benthic host species that tend to sweep the substrate and juvenile mussels out of the cage (e.g. catfish), a small "spacer cage" can be added to keep the fish off the bottom (Figure 6.11e). The juveniles fall into the sandy substrate inside the spacer cage while the fish remain in the upper cage.

Maintenance of the mussel cages is critical. After deployment in a river for two summers, cage screening is typically degraded and in need of replacement. The plywood base can crack and warp over time and should be inspected and replaced if necessary. Welds in the cage

(a)

(b)

FIGURE 6.11 (a) A cage used for outdoor mussel culture at the Genoa National Fish Hatchery (Genoa, WI). (b) Juvenile freshwater mussels being added to a cage for grow-out. A fine layer of sand is added to the bottom of the cage to allow the mussels to burrow. (c) An underwater photograph of the standing cages at Genoa National Fish Hatchery (Genoa, WI). In this case, the metal legs keep the cage off the bottom of the pond. (d) Outdoor mussel culture cages on floating racks at Genoa National Fish Hatchery (Genoa, WI). The white flotation buoy in the upper right hand corner of the photo is used to keep the cages off the bottom of the pond. (e) Nathan Eckert, USFWS, inspecting a spacer cage at Genoa National Fish Hatchery (Genoa, WI). The large cage on top is being lifted to reveal the smaller cage below. Juvenile mussels drop off the host and fall into the lower cage where they are physically separated from the fish above and less likely to get swept from the cage.
Photos: Ryan Hagerty, USFWS. (A black-and-white version of part (d) of this figure will appear in some formats. For the color version, please refer to the plate section.)

(c)

(d)

(e)

FIGURE 6.11 (cont.)

frame also may crack and require maintenance. For cages that only spend part of the year submerged (i.e. May–October), the expected lifespan may be extended by an additional season or two.

6.4.2 Floating Baskets

The Virginia Fisheries and Aquatic Wildlife Center at Harrison Lake National Fish Hatchery (Charles City, Virginia) developed a floating basket that has been adapted by several other mussel propagation facilities (Figure 6.12a). The floating basket system consists of a fish basket, 150 micron nylon mesh, 200 micron nylon mesh, play sand and sediment, large and small pool noodles, 8 and 14 inch cable ties, and plastic mesh for a lid. The 150 micron nylon mesh is glued to the bottom of the basket using silicone. The silicon seal holding the nylon mesh can separate after one season and may need repair. Stainless steel bolts with large nylon washers can be used to secure the nylon mesh in place for multiple years of use (Figure 6.12b). A second, oversized piece of nylon mesh (typically 200 micron) is placed on top of the 150 micron nylon mesh (Figure 6.12c). About 1–2 inches of rinsed, fine play sand and 500 mL of sediment are added and the juvenile mussels are placed on top (Figure 6.12d). To ensure the sand and the juvenile mussels do not get washed out through the holes on the side of the basket, the oversized piece of 200 micron nylon screen should come up the sides of the basket about 1–2 inches. Large and small pool noodles are attached to the basket with 14 inch cable ties to provide flotation (Figure 6.12e). Plastic mesh screen covers the top, attached with 11 inch cable ties.

Floating baskets are deployed when water temperatures begin to warm in the spring. The basket filled with sand and large mussels can be easily moved by one person, making them relatively easy to deploy and sample. To place juvenile mussels in a floating basket, a 3 inch piece of PVC is inserted vertically inside the basket with one end sticking out of the water. The juvenile mussels are slowly poured through the PVC and given time (usually 5–10 minutes) to settle to the bottom of the basket. The PVC tube is removed slowly and a plastic mesh lid is attached. Juvenile mussels of any size and

(a)

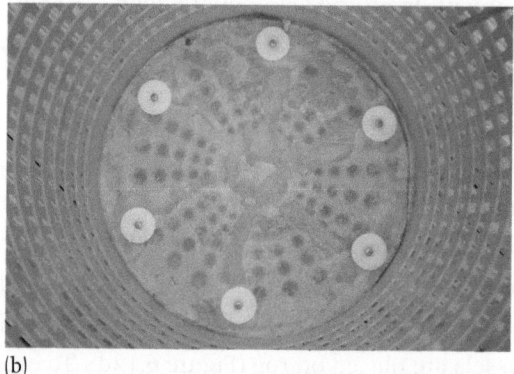

(b)

FIGURE 6.12 (a) A freshwater mussel culture basket developed at
the Virginia Fisheries and Aquatic Wildlife Center at Harrison Lake
National Fish Hatchery (Charles City, VA). The pool noodles are
attached to the basket to provide flotation. (b) The inside of the basket
showing the 150 micron mesh and stainless steel bolts and large nylon
washers used to keep the mesh in place. (c) An oversized piece of 200
micron mesh is placed on top of the 150 micron mesh. (d) A fine layer
of sand is placed on the 200 micron mesh to allow the mussels to
burrow. (e) Baskets floating on the ponds at Harrison Lake National
Fish Hatchery (Charles City, VA). A plastic mesh lid is attached to
the top of the basket to exclude potential predators and a waterproof
tag is added to identify the species and lot number in the basket. (f)
Subadult eastern pondmussel (*Ligumia nasuta*) and yellow lampmussel
(*Lampsilis cariosa*) inside a floating basket at the Virginia Fisheries
and Aquatic Wildlife Center at Harrison Lake National Fish Hatchery
(Charles City, VA).
Photos: (a, b, c, d, e) Ryan Hagerty, USFWS. (f) Matthew Patterson,
USFWS. (A black-and-white version of part (f) of this figure will appear
in some formats. For the color version, please refer to the plate section.)

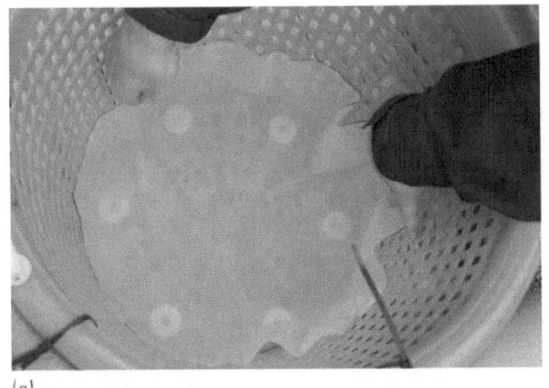

(c)

(d)

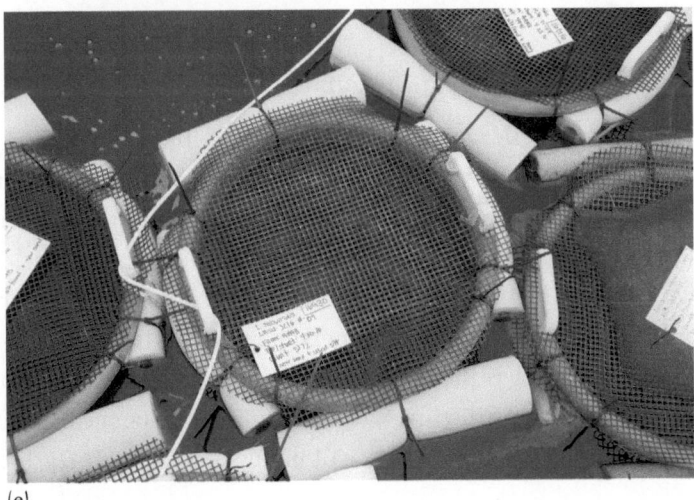

(e)

FIGURE 6.12 *(cont.)*

(f)

FIGURE 6.12 *(cont.)*

age can be deployed in floating baskets; however, survival rates for most species are higher when they are 3–10 mm in length. Mussels are typically removed from the cages within 3–18 months, tagged, and released (Figure 6.12f).

After 1 year in a pond, maintenance is usually required to get the baskets ready for the following season. Baskets should be power-washed or scrubbed and the nylon mesh screen and pool noodle flotation may need replacement. The sand can be washed and saved for future years.

6.4.3 Floating Upweller System (FLUPSY)

The floating upweller system (FLUPSY) was first designed by the marine bivalve industry. When it became clear that indoor culture systems could not supply enough food, even using flow-through, wild-water sources, they moved to outdoor culture systems. The FLUPSY uses a floating raft with holes cut in the bottom to hold juvenile culture chambers. A submersible pump drives water out of the floating raft and as the raft refills, water and food upwell through the culture chambers (Figure 6.13a). Culture chambers can be constructed using 6–12 inch PVC, 5 gallon buckets, or even trash cans with 0.8–1.5 mm mesh screens glued to the bottom. The FLUPSYs designed by Missouri State University and housed at the Kansas City Zoo use plastic trash cans hanging from a dock and floating in the Zoo's lagoon (Figure 6.13b, c). Typically, juvenile mussels greater than 3–5 mm in length are deployed in FLUPSYs. Depending on the mesh size of the culture chamber and organic content of the pond or other water source, screens may clog and require daily or weekly cleaning.

6.4.4 Suspended Upweller System (SUPSY)

The suspended upweller system (SUPSY) was designed by the Alabama Department of Conservation and Natural Resources' Alabama Aquatic Biodiversity Center (Marion, Alabama). The SUPSY also utilizes upwelling to deliver food to the juvenile mussels; however, this system is suspended in the water column instead of floating on the surface. The system includes a pair of nested two-gallon buckets with a mesh screen on the bottom of both buckets (Figure 6.14a). Fiberglass reinforced pet screening works best because it is more flexible and durable than standard window screen or nylon screen (Figure 6.14b, c). Water flow is driven by 1.0 inch air lift tubing inserted in the lid of the upper bucket (Figure 6.14d). A single regenerative blower can generate upwelling water flow in several SUPSYs at the same time. A brick is attached to the bottom bucket to keep the system anchored, while small pieces of pool noodle, or net floats, keep the system upright and off the bottom. A full-size brick is typically required to offset the buoyancy of the two plastic buckets. These systems can be set close to the bottom or high in the water column, depending on local conditions at the deployment site. Lengthening the anchor line or tying the SUPSY to an anchor line from the surface will change its position in the water column (Figure 6.14e). Screens should be inspected and cleaned every 1–2 weeks, depending on the amount of sedimentation and debris at the deployment location. The SUPSY provides a very effective and low-cost method for growing juvenile mussels greater than 3 mm in length up to a size suitable for tagging.

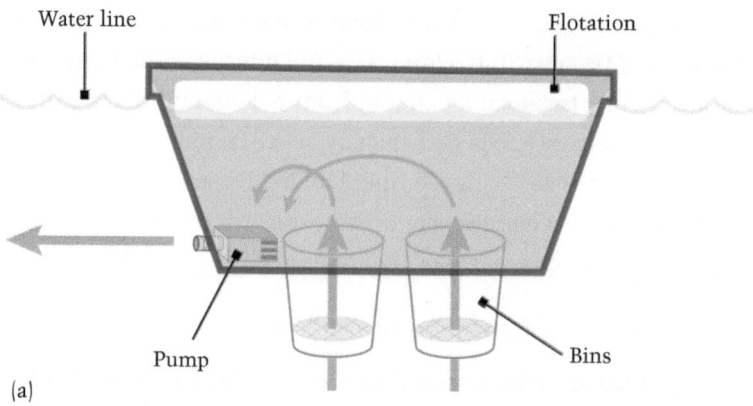

Water line   Flotation

Pump   Bins

(a)

(b)

FIGURE 6.13 (a) A diagram showing the design and water movement in a FLUPSY. A submersible pump bails water out of the main tank and forces new water up through the culture chambers. (b) A FLUPSY designed by Missouri State University and housed at the Kansas City Zoo showing the floating dock and trash can culture chambers suspended below. (c) A close-up of the FLUPSY culture chambers at the Kansas City Zoo.
Diagram: (a) Courtesy of Dr. Chris Barnhart (Missouri State University) and adapted by Kristin Simanek, USFWS. Photos: (b, c) Chris Barnhart, Missouri State University.

(c)

FIGURE 6.13 (cont.)

6.4.5 Mussel Silos and Bunkers

Mussel silos, developed by Dr. Chris Barnhart at Missouri State University, are dome-shaped concrete structures with a mesh screen culture chamber inside to house the juvenile mussels (Figure 6.15a–d). The heavy concrete base and low profile allows caged mussels to be deployed into high-flow areas of rivers and streams (Figure 6.15e, f). As water flows over the top of the dome, water under the dome is forced up through the culture chambers (upwelling) much like air passing over an airplane wing causes upward lift (i.e. the Bernoulli effect). The upwelling flow delivers food and oxygen to the juvenile mussels inside the culture chamber. Growth rates are limited by food availability and temperature at the release site.

Silos have been used for juvenile grow-out and *in situ* contaminant studies, as well as temporary holding for imperiled brood stock that may be brooding unfertilized eggs. Silos also can be useful for

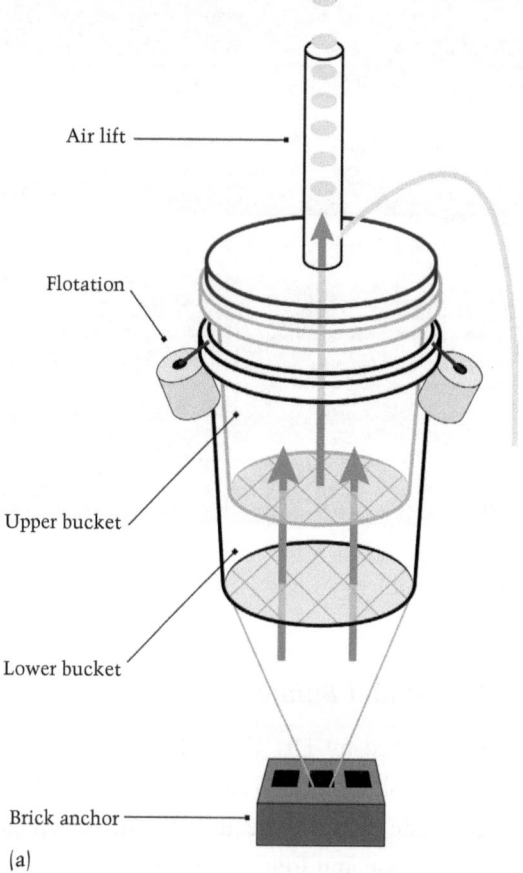

Air lift

Flotation

Upper bucket

Lower bucket

Brick anchor

(a)

FIGURE 6.14 (a) Diagram of the original SUPSY design at the Alabama Aquatic Biodiversity Center (Marion, Alabama) showing the direction of water flow. An air lift bails water out of the bucket, forcing water to upwell through the culture container. The brick anchor keeps the SUPSY in place while flotation keeps the system off the bottom of the pond. (b) Inside of a SUPSY culture chamber showing the mesh screen that holds the juvenile mussels. (c) The inside of a SUPSY culture chamber with juvenile mussels resting on the mesh screen. (d) Top of a SUPSY showing the PVC airlift tube and air nozzle. (e) A suspended upwelling system (SUPSY) being deployed from a pier at Genoa National Fish Hatchery (Genoa, WI).

Diagram: (a) Courtesy of Paul Johnson (Alabama Aquatic Biodiversity Center) and adapted by Kristin Simanek, USFWS. Photos: (b, d) Matthew Patterson, USFWS. (c, e) Ryan Hagerty, USFWS. (A black-and-white version of part (c) of this figure will appear in some formats. For the color version, please refer to the plate section.)

(b)

(c)

(e)

(d)

FIGURE 6.14 *(cont.)*

(a)

FIGURE 6.15 (a) A mussel silo designed by Dr. Chris Barnhart showing the concrete dome and mussel culture chamber in place. (b) The mussel culture chamber removed from the mussel silo. (c) The mussel culture chamber taken apart to show the mesh screens. (d) The mussel silo culture chamber with juvenile mussels on the mesh screen. (e) A mussel silo deployed in the Red Bird River (Clay County, KY). The silo was deployed by the United States Forest Service in partnership with the Kentucky Department of Fish and Wildlife Resources and the Kentucky Division of Water. (f) A mussel silo deployed in the river with a piece of tubing inserted in the mussel culture chamber to collect water samples for analysis.

Photos: (a) Julie Campbell, USFWS. (b, c, d, f) Amy Thompson, USFWS. (e) Wendell Haag, US Forest Service.

determining the suitability of potential release sites (see Chapter 7). Mussel bunkers, also developed by Chris Barnhart at Missouri State University, are a larger version of the mussel silos and can be used to hold large numbers of juvenile mussels or adult brood stock in rivers (Figure 6.16). Deployment sites for mussel silos and mussel bunkers should be carefully selected because vandalism, screen fouling, and

(b)

(c)

FIGURE 6.15 (cont.)

(d)

(e)

FIGURE 6.15 (cont.)

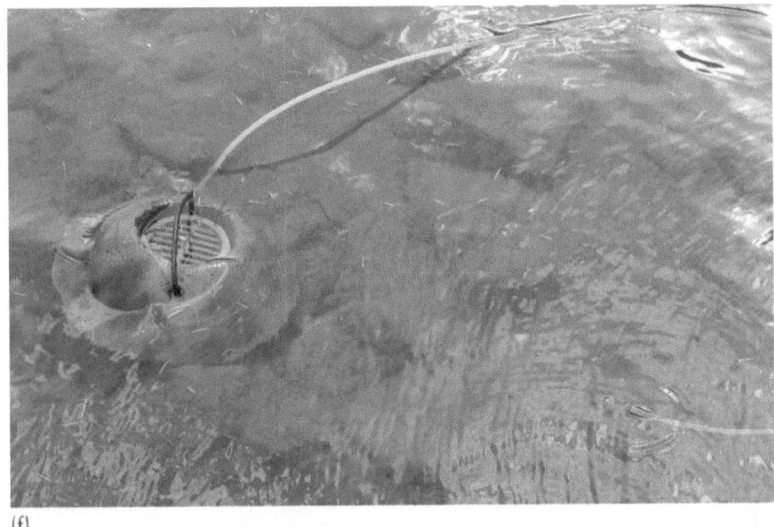

(f)

FIGURE 6.15 (cont.)

(a)

FIGURE 6.16 (a) A mussel bunker designed by Dr. Chris Barnhart at Missouri State University. The dome shape of the bunker creates upwelling flow as described in the mussel silos. (b) One of the mussel culture chambers being removed from the mussel bunker. (c) A close-up photograph of the mussel culture chamber from the mussel bunker. (d) The mussel culture chambers taken apart for demonstration. These chambers can hold subadult mussels or adult brood stock.
Photos: Chris Barnhart, Missouri State University.

(b)

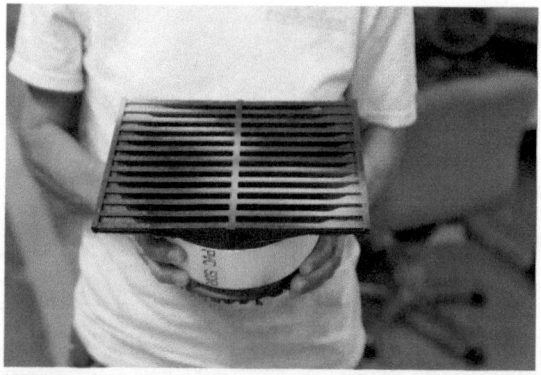

(c)

(d)

FIGURE 6.16 (*cont.*)

sediment trapping can effect juvenile recovery and survival. Painting silos and bunkers to blend in with the surroundings or deploying them in shady areas of the stream, amongst large substrate, or downstream of logs can help prevent vandalism.

6.4.6 Streamside Culture

A mobile rearing unit (MRU) such as the Mobile Aquatic Rearing System (MARS) at the Genoa National Fish Hatchery (Genoa, WI) can be used to culture juvenile mussels along the side of a stream (Figure 6.17a). An MRU can be useful if water quality problems, biosecurity issues, or space are limiting factors at an existing facility. The primary benefit of these units is the ability to pump wild water from a river or other body of water known to support native freshwater mussels. A wild-water source helps ensure the water quality and dietary requirements of the juvenile mussels are being met. An MRU also can be set up to draw water from a hatchery pond if conditions are favorable. Other benefits include culturing freshwater mussels off-site, reducing the space and infrastructure needs associated with a traditional propagation facility, and avoiding any biosecurity risks for existing fish culture operations.

The MARS is a 20 foot cargo trailer outfitted with pumps and filtration for flow-through mussel culture. A submersible trash pump enclosed within an intake screen pumps water through a drum filter and ultraviolet (UV) sterilizer before entering the culture tanks. The fiberglass culture tanks are held inside rectangular, aluminum raceways and have successfully been used to culture both newly metamorphosed juveniles and subadult freshwater mussels (Figure 6.17b). A second UV system sterilizes the outflow water before it returns to the river.

6.5 FEEDING AND NUTRITION

Regardless of the culture system chosen, it is critical to provide juvenile mussels with adequate nutrition for both growth and survival. Unfortunately, very little is known about the nutritional requirements

(a)

(b)

FIGURE 6.17 (a) The Mobile Aquatic Rearing System (MARS) at the Genoa National Fish Hatchery. (b) The interior of the MARS at the Genoa National Fish Hatchery showing the aluminum raceways and fiberglass mussel culture tanks.
Photos: Ryan Hagerty, USFWS.

of freshwater mussels. Most of the available information comes from the marine bivalve literature. While specific nutritional requirements for juvenile freshwater mussels are relatively unknown, it is becoming clear that successful culture of juvenile mussels inside or outside the laboratory depends on providing a balanced diet. The following recommendations should be considered when developing a diet to feed juvenile mussels.

6.5.1 Feed a Diverse Diet

A freshwater mussel culture program must provide a balanced diet with adequate levels of protein, carbohydrate, and lipid to be successful. These nutrients are the most important biochemical constituents of a suitable bivalve diet. Protein is responsible for tissue production and is essential for juvenile growth. Protein-limited diets have been shown to decrease growth in the bay mussel, *Mytilus trossulus* (Kreeger and Langdon, 1993). Carbohydrates, primarily glycogen, are the primary energy source for bivalves. Dietary carbohydrates also may be important in balancing proteins and lipids for energy production (Whyte *et al.*, 1989). Enright *et al.* (1986) showed that a diet high in carbohydrates increased growth of European flat oyster larvae (*Ostrea edulis*), if adequate levels of protein and lipid were available. Lipids are important in reproductive development and likely are a major energy source for the developing larvae (Berenberg and Patterson, 1981). It is unlikely that lipids are synthesized by freshwater mussels and therefore must be available in the diet.

Algae can be high in lipids or plant sterols and are therefore important components of any bivalve diet. Juvenile freshwater mussels have been reared successfully on algal diets containing high levels of fatty acids (Gatenby, 1994; Gatenby *et al.*, 1997, 2003). Microalgae species differ in their biochemical components, so a multi-species diet will be more balanced. A combination of several microalgae species, including diatoms, has been shown to increase growth and survival for bivalves (Brown *et al.*, 1997; Gatenby *et al.*, 2003; Helm and Bourne, 2004; Mair, 2013). Additionally, live algal cells and algae mixes have been shown to increase bivalve growth rates compared to dead algal cells (Gatenby *et al.*, 1996, 1997; Heasman *et al.*, 2001; Mair, 2013). Although live algal cells are preferred, the costs of culturing algae on-site can outweigh the benefits, especially if feeding live diets results in marginal increases in growth (less than 1 mm) and survival (Mair, 2013). Juvenile mussels started out on dead algal cells quickly make up any growth deficit when exposed to a natural, wild-water source in secondary culture.

Recent research in freshwater mussel propagation seems to indicate that the best way to provide a balanced diet is to provide some sort of "wild water." In some cases, growth rates are orders of magnitude faster when mussels are fed or supplemented with wild water (Figure 6.18). River water, pond water, and other wild-water sources contain an enormous diversity of algae, bacteria, and detritus that is nearly impossible to replicate in the laboratory using cultured food. The diversity of food resources available in wild water is the likely cause for increased growth rates. The downside of wild-water sources is the composition of the diet can change seasonally and the water source can be susceptible to contamination, invasive species, predators, sedimentation, or toxic materials.

6.5.2 Feed Small Cells to Young Juveniles

The size of the food particle is an important consideration when developing a diet for juvenile freshwater mussels. In the wild, mussels ingest particles less than 28 microns in diameter, including algae, detritus, and bacteria (Nichols and Garling, 2000). Lasee (1991) determined a 2-day-old juvenile has a mouth of about 16 microns and an esophagus of about 6 microns. Beck and Neves (2003) determined that smaller cells (2.8–8.5 microns) are preferred to larger cells (22.8–44.5 microns), which are passed over. To ensure proper ingestion, young juveniles less than approximately 1 mm in length should be fed particles 1–10 microns in diameter.

FIGURE 6.18 Four-month-old alewife floater (*Anodonta implicata*) cultured in floating baskets on wild pond water by the Virginia Fisheries and Aquatic Wildlife Center at Harrison Lake National Fish Hatchery.
Photo: Rachel Mair, USFWS. (A black-and-white version of this figure will appear in some formats. For the color version, please refer to the plate section.)

6.5.3 Do Not Over-feed

Both absorption efficiency and filtration rate of juvenile freshwater mussels can be negatively affected by high feed concentrations (Bush, 2008). Adult oyster mussel (*Epioblasma capsaeformis*), Cumberlandian combshell (*Epioblasma brevidens*), and snuffbox (*Epioblasma triquetra*) fed a low concentration of algae (20 000 cells per mL) had significantly higher absorption efficiencies than medium (40 000 cells per mL), high (80 000 cells per mL), and very high (120 000 cells per mL) feed rations (Bush, 2008). For information on known filtration rates of freshwater mussels, refer to Gatenby *et al.* (2013). For the striated fingernail clam (*Sphaerium striatinum*), filtration rate decreased as particle concentration increased (Hornbach *et al.*, 1984). Numerous marine bivalve studies have investigated the effect of food concentration on filtration rate. In these studies, filtration rate increased until a maximum amount of food was ingested (Winter, 1978; Navarro and Winter, 1982). Once the maximum ingestion rate was reached, filtration rate decreased until the production of pseudofeces began. Further increases in feed concentration led to reduced rates of filtration and ingestion (Winter, 1978; Navarro and Winter, 1982). High feed rates also can lead to the accumulation of ammonia in recirculating systems.

Early freshwater mussel propagation research indicated that a feed ration of 20 000–30 000 cells per mL was sufficient for juveniles; however, other studies have shown excellent survival at much higher densities, 100 000–500 000 cells per mL (Gatenby *et al.*, 1997, 2013; Rodgers, 1999; Henley *et al.*, 2001). The concentration of food needed will depend on the culture system being used and the amount of weekly maintenance being performed. The mussel species in culture and the quality of feed also will affect the optimal feed ration. Northern riffleshell (*Epioblasma torulosa rangiana*) and mucket (*Actinonaias ligamentina*) cultured in bucket systems at three feed concentrations (30 000, 90 000, and 150 000 cells per mL) showed the highest growth and survival ($p < 0.01$) at the lowest feed concentration (Mair, 2013). Consequently, the recommended starting concentration for juvenile mussels is 30 000–50 000 cells per mL (or a volume of 1–3 × 10⁶ cubic microns per mL). Measuring cell concentration in cubic microns per mL is preferred over cells per mL because it takes into account the size of the algal cell; thus, providing a better estimate of the dry weight of food available. The starting concentration can always be adjusted based on consistent monitoring of feed concentrations in the culture containers, as well as growth and survival of the juveniles. Consistent monitoring and detailed record-keeping will help identify the ideal feed concentration for any given mussel species and culture system combination. Feed concentration can be monitored using a hemacytometer or cell counter (Figure 6.19). Visual inspections of the mussel culture containers should reveal a very light, but noticeable tinge of green in the water.

6.5.4 Live or Commercial Diets

Algae that is cultured on-site contains mostly live cells, while commercially available algae feed concentrates are composed of dead cells. Sydney rock oysters (*Saccostrea glomerata*) fed fresh algae had twice the growth of oysters fed algae with additives (Heasman *et al.*, 2001). Ponis *et al.* (2003) showed a decrease in growth for juvenile Pacific oysters (*Crassostrea gigas*) when fed concentrated diets. Survival during these experiments was not significantly different when oysters were fed live algae compared to centrifuged or preserved algae. Other marine studies have compared growth and survival among microalgae concentrates and reported that, generally, as long as the concentrates were not old (less than weeks), at least partial substitution of concentrates for live algae did not yield detrimental results (Aji, 2011). Mair (2013) found significantly higher growth rates for juvenile northern riffleshell (*Epioblasma torulosa rangiana*) and mucket (*Actinonaias ligamentina*) fed live algae (3.1 mm) compared to two commercial diets (2.6, 2.4 mm). No difference in survival was documented between the commercial diet and live algae mix.

(a)

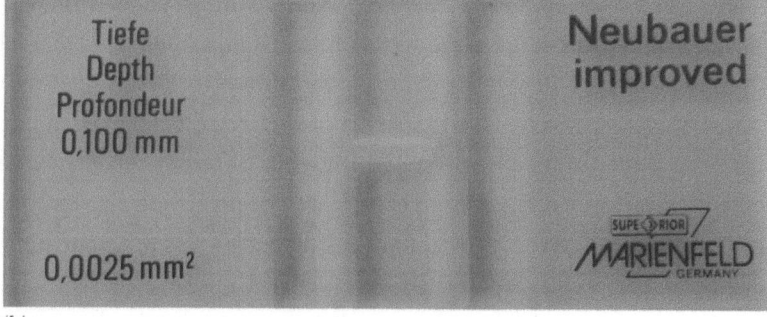

(b)

FIGURE 6.19 (a) The Beckman Coulter Multisizer cell counter at Virginia Fisheries and Aquatic Wildlife Center at Harrison Lake National Fish Hatchery. The multisizer can be used to track feed concentrations in the mussel culture chambers. (b) A hemacytometer is a less expensive but more labor-intensive option for tracking feed concentration. (c) A close-up of the hemacytometer grid used to enumerate algal cells.

Photos: (a) Ryan Hagerty, USFWS. (b, c) Rachel Mair, USFWS.

(c)

FIGURE 6.19 (*cont.*)

If a propagation facility does not have access to a wild food source or if wild water poses a biosecurity risk, then culturing live algae on-site may be necessary, supplemented with commercial algae concentrates. A controlled indoor environment is recommended for culturing multiple species of algae, which will increase operational costs for the facility (Southgate *et al.*, 1992; Heasman *et al.*, 2001). The aquaculture industry, however, also produces algae outdoors in ponds. Production costs for algae culture in the United States range from $160–400 per kg, which accounts for almost 30% of marine production costs (Southgate *et al.*, 1992; Knauer and Southgate, 1999; Heasman *et al.*, 2001).

Algal cultures (marine and freshwater) will experience occasional and unpredictable crashes for a variety of reasons and these culture crashes may occur at crucial production times. To reduce operational costs and potentially devastating results from algae culture crashes, artificial diets continue to be tested for their commercial value in the marine bivalve industry (Knauer and Southgate, 1999; Heasman *et al.*, 2001). For freshwater mussels, recent successes achieved with natural food sources (i.e. pond or river water) could eliminate the need for cultured or purchased algal diets. Most propagation facilities use river or pond water supplemented with either cultured algae or commercial diets. If live algae are needed, Appendix E provides a manual and instructional video link for culturing algae on-site.

6.5.5 Automated Feeding Systems

It is critical that juvenile and adult mussels be fed continuously (24 hours a day) in the laboratory (see Section 4.5). Several options are available for automatic feed delivery including timers and solenoids, dosing pumps, intravenous (IV) drip bags, and peristaltic pumps. Most automated feeding systems use timers and solenoid valves to deliver food continuously (Figure 6.20). Each solenoid valve is connected to a timer that controls when the solenoid opens and distributes algae to the culture chamber. The length of time the solenoid valve is open controls the amount of algae added.

6.6 ASSESSING GROWTH AND SURVIVAL

To evaluate the success of the propagation program, juvenile growth and survival must be monitored regularly. Regular monitoring will allow results to be statistically analyzed and compared with other facilities. Shell length measurements should be collected every 2–4 weeks, depending on available staffing. Sampling intervals should remain consistent to allow results to be easily compared over time. Record length measurements and juvenile counts for each culture chamber, especially if juveniles are being moved from one culture system to another. Research indicates that handling stress can negatively affect growth and possibly survival, so be careful not to over-sample.

Shell length is measured as the greatest distance between the posterior shell margin and the anterior shell margin using calipers

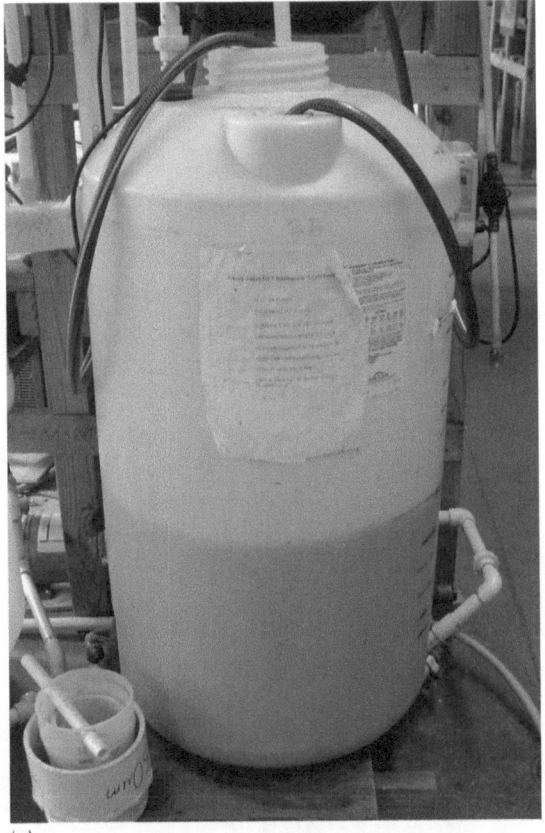

(a)

FIGURE 6.20 (a) The bulk algae feed tank at White Sulphur Springs National Fish Hatchery (White Sulphur Springs, WV). Concentrated algae are continuously pumped through a PVC manifold and back to the bulk tank. (b) A series of repeat cycle timers are used to control the feed frequency and feed amount delivered to the culture systems. (c) Solenoid valves are plumbed into the PVC manifold and open and close in response to the settings on the repeat cycle timers. When the valves are open, algae is dispensed into the culture container. (d) At Virginia Fisheries and Aquatic Wildlife Center at Harrison Lake National Fish Hatchery (Charles City, VA), each juvenile culture system has its own bulk algae container (nalgene bottle), solenoid valve, and timer. (e) A close-up photograph of the bulk algae containers at Virginia Fisheries and Aquatic Wildlife Center at Harrison Lake National Fish Hatchery (Charles City, VA).
Photos: Matthew Patterson, USFWS.

(b)

(c)

(d)

FIGURE 6.20 (*cont.*)

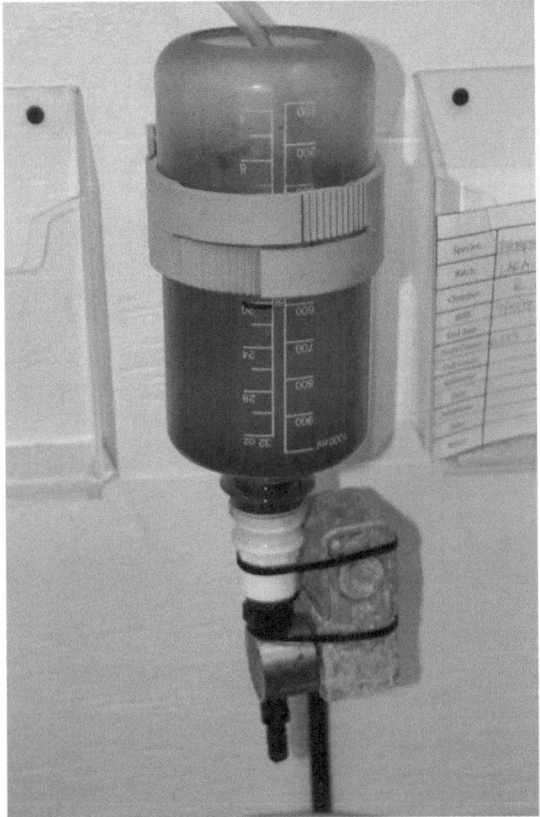

(e)

FIGURE 6.20 (cont.)

(Figure 6.21a). Photography and image analysis software also can be used to assess growth of small juveniles. Shell length is typically the only measurement taken, however, shell height and shell width also can be recorded (Figure 6.21b, c). During each sampling event, measure a minimum of 20 mussels from each culture container. The ideal sample number may change depending on the size of the culture chamber and the number of mussels present. When conducting culture experiments, increasing the number of individuals measured will lower the statistical variance and strengthen the data analysis.

(a)

(b)

(c)

FIGURE 6.21 (a) Shell length is measured from the anterior shell margin to posterior shell margin. (b) Shell height is measured from the dorsal shell margin to the ventral shell margin. (c) Shell width is measured from the left valve margin to the right valve margin.
Photos: Rachel Mair, USFWS. Graphics: Kristin Simanek, USFWS.

6.7 CONTROLLING UNWANTED ORGANISMS
IN THE CULTURE ENVIRONMENT

Some of the most common unwanted organisms that can show up in mussel culture systems include flatworms, chironomids, ciliates, daphnids, and others. Little information is available on how to control unwanted organisms in the culture environment without causing harm to the juvenile mussels, so prevention is the best option. To prevent an outbreak of unwanted organisms, it is important to maintain clean culture systems with very high water quality. Mechanical, biological, ultraviolet, and/or chemical filtration methods should be used (when possible) on all juvenile mussel culture systems. Mechanical filters like cartridge filters, sponge filters, and floss remove particulate debris from the water. Sponge filter media can induce "blooms" of Cladocera (daphnids) and should be cleaned regularly. Biological filters, including sand, fluidized bed, and trickle filters, as well as filter media that increase surface area for bacterial colonies, break down waste and convert harmful ammonia and nitrogen into non-toxic forms. Chemical filtration, including activated carbon filters are inexpensive and can be low maintenance, but are not always necessary to control unwanted organisms. Proper filtration on all fish systems is also important as many unwanted organisms are transferred from the fish host systems. Some organisms also can simply enter through the air, so keep culture systems as closed as possible. Following basic quarantine and biosecurity protocols like keeping nets, siphons, and maintenance cleaning supplies separate for each system, disinfecting with bleach or other disinfectant before using equipment on a different system, and keeping all equipment off of the floor can help prevent the spread of unwanted organisms.

If an unwanted organism outbreak occurs in a culture system, there are a couple of control options. Fungal outbreaks, for example, can be controlled using ultraviolet sterilization, frequent water changes, or by increasing the temperature to 28–34 °C for 24 hours. Most larvae and small invertebrates will not survive exposure to

FIGURE 6.22 Gnatrol WGD, a biological larvicide that can be applied to control dipteran larvae in juvenile culture systems.
Photo: Rachel Mair, USFWS.

ultraviolet light and some mussel species will likely be unable to handle the increase in temperature. One effective method for controlling insect infestations, including chironomids, is the use of biological larvicides, or products similar to Gnatrol WDG (Figure 6.22). Several treatments of 10 ppm Gnatrol has been effective against reducing dipteran larvae without harming the juvenile mussels.

7 Juvenile Mussel Release and Monitoring

Bryan R. Simmons, Matthew A. Patterson, and Jess W. Jones

Releasing juvenile freshwater mussels to the wild and subsequent monitoring should be considered the most important step in the entire mussel propagation process. Significant amounts of time, energy, staff, and funding have been invested in these juvenile mussels, so the utmost care should be taken when releasing them to the wild. Once the mussels have been released, evaluating the success or failure of the propagation program will be difficult to assess if the mussel population at the release site is not monitored.

The logistics of releasing juvenile freshwater mussels to the wild can be very complex. Several considerations should be taken into account before release, including the purpose or goal of the release, site suitability, site conditions, disease prevention, effective tagging, proper transportation, stocking size, stocking density, staff support, release technique, and record-keeping. Many of these considerations (i.e. release purpose, recovery goals, and release site suitability) should be addressed during the propagation planning process, well before adult brood stock or the host species are brought on station (see Chapter 1). Finally, make sure you have the appropriate state and federal permits to possess and transport freshwater mussels to prevent any potential violations, especially if you are transporting mussels across state lines. If the goals and objectives were clearly defined and suitable release sites were identified, then more often than not releasing mussels to the wild will be a low stress and successful event.

Any plan to propagate and release juvenile freshwater mussels also must include a plan for long-term monitoring. This step oftentimes falls by the wayside due to limited time, staffing, and budgets, but monitoring is essential for adjusting conservation strategies in an "adaptive management" framework. Ultimately, the monitoring plan should track progress toward the conservation objectives identified in the Propagation Plan (or species recovery plan). Key elements of a basic monitoring plan should include monitoring goals and objectives, geographic scope of the monitoring area, duration and frequency of monitoring, sampling design, universal metrics of measurement, and strategies for ensuring monitoring data are fed back into the propagation planning process (Atkinson *et al.*, 2004).

While the authors feel strongly that post-release monitoring is a key step in the mussel propagation process, the primary purpose of this book is to cover the basic methods for raising freshwater mussels in a propagation facility. Consequently, this chapter will provide a broad overview of mussel monitoring and the importance of feeding monitoring data back into the propagation planning process. When planning a post-release monitoring program, please consult the primary literature and books regarding standard protocols for monitoring freshwater mussels (e.g. Strayer and Smith, 2003).

7.1 JUVENILE RELEASE

7.1.1 *Tagging*

Prior to juvenile release, a tag or other identifying mark should be placed on the shell for future monitoring (Figure 7.1). Tagging allows for assessment of the program's success or failure by differentiating between propagated mussels and natural recruitment. Tagged mussels also provide exceptional models for various ecological, evolutionary, and toxicological studies of long-lived animals. In such studies, external identification using tags or engravings are the primary means of tracking growth, survivorship, longevity, and movement.

For a tag to be effective, it should be inexpensive and easy to apply to a large number of mussels, permanently attached to the shell and legible over the duration of the study or monitoring protocol. These criteria can be difficult to meet because freshwater mussels tend to live in very abrasive environments and can be very long-lived. Tags can be exposed to a variety of rough conditions in the environment,

(a)

(b)

FIGURE 7.1 (a) A SCUBA diver releasing tagged, subadult northern
riffleshell (*Epioblasma torulosa rangiana*) to the Allegheny River.
(b) Subadult fatmucket (*Lampsilis siliquoidea*) tagged with Hallprint
tags and transported to the release site in a mesh bag.
Photos: (a) Janet Butler, USFWS. (b) Rachel Mair, USFWS. (A black-and-
white version of this figure will appear in some formats. For the color
version, please refer to the plate section.)

including abrasive substrates (gravel and sand) and chemical erosion forces. It is important to pay close attention to where a tag is placed on the shell. The umbo is typically the most eroded feature of a mussel shell. Tag retention will increase if tags are placed near the center of the shell, away from this area of high erosion.

Tags can be as simple as a dot of colored superglue or as complex as a ten character-number set, laser engraved into the shell. Simple tags consist of simple marks that cannot be confused with natural erosion. Complex tags require special adhesion or application techniques and usually consist of unique letter and/or number combinations. All tag types have advantages and disadvantages based on cost, ease of application, longevity, interpretation, and duration. The ideal will depend on monitoring goals, duration of the monitoring project, mussel species, and water chemistry.

Shell engraving is the most permanent method for tagging mussel shells. Tools for shell engraving range from basic knives or files to Dremel tools and laser engraving machines. When engraving a shell with a knife or file, simply remove the periostracum and one level of the nacreous layer, leaving a visual mark. Simple symbols such as X, O, Δ, or numbers should be used (Figure 7.2a). Legibility is a primary concern for simple etched markings and could lead to errors in data collection and interpretation. Try to avoid using complex symbols and multiple similar markings for different groups of mussels.

Shells also can be engraved with a Dremel tool outfitted with a 3/32 to 1/8 inch spherical burr bit (Figure 7.2b). Select an area of the shell as far away from the umbo as practical, where the periostracum is in good condition. The cut from the Dremel tool should be deep enough to enter the nacre but not so deep as to perforate the shell. Larger marks are easier to read and last longer so make the mark as large as possible depending on the size of the shell (up to about 1 cm or even 2 cm tall on large mussels). Engraved marks on adult shells can remain visible for decades. It may not be feasible to etch both the left and right valves, especially if a large number of mussels will be marked. If smaller numbers of mussels will be etched

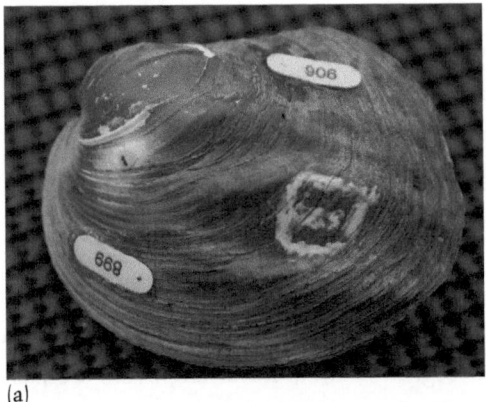

(a)

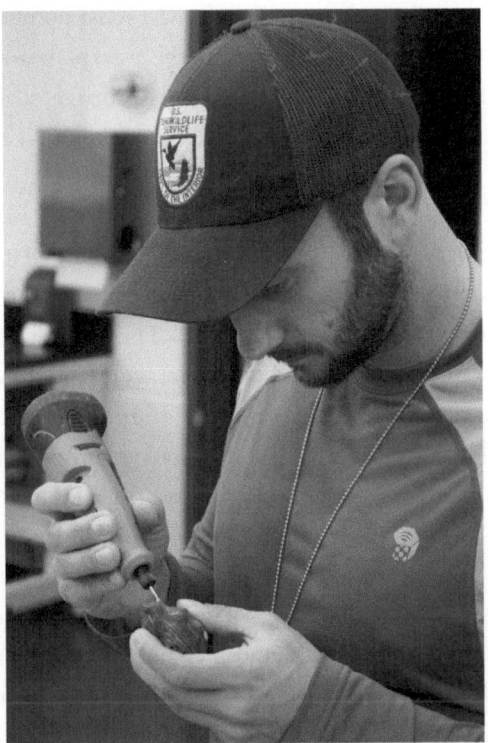

(b)

FIGURE 7.2 (a) A threeridge (*Amblema plicata*) with several tag types including numbered glue tags and a simple engraved tag. (b) USFWS personnel using a Dremel tool to engrave a tag into a freshwater mussel shell.
Photos: Matthew Patterson, USFWS.

or time is not a concern, it is better to etch both valves. The two valves can easily get separated when a mussel dies and identifying the shell as part of a monitoring project will be impossible if the unetched valve is collected.

Manually engraving a large group of mussels with a knife or Dremel tool can be time-consuming. Missouri State University began using a laser engraver, originally designed to customize commercial products like signs, name tags, etc., to engrave a unique number-letter combination onto the shell of a large number of mussels in a short period of time (Figure 7.3a, b). Laser engravers are basically flat-bed dot-matrix printers that can be adjusted to burn a groove in the shell, one dot at a time. A guide is used to fix the focal point of the laser in a particular plane so the mussels must be relatively uniform in size and shape (Figure 7.3c). Subadult mussels propagated in the lab are typically the same age and grown under similar conditions, so a laser engraver can be an excellent time-saving option. Heating is localized and the speed of printing (dots per minute) must be adjusted so that the mantle underneath the shell does not overheat. The mussels are arrayed on a grid pattern and an Excel spreadsheet is used to position the marks on the grid (Figure 7.3d, e, f). The staff at Missouri State University use an Epilog Zing-24 laser engraver (Figure 7.3g) and working alone, one person can engrave several hundred mussels per hour.

While shell engraving can be relatively inexpensive, it can encourage shell erosion over time, especially in soft-water environments. Shell engravings also may not be suitable for tagging small juveniles or large juveniles of thin-shelled mussel species.

Glue tags and glue-on tags use epoxies or cyanoacrylate for underwater application. The glue can either be used alone to create a mark or to adhere a manufactured tag to the shell. Tag types include glue dots, Hallprint tags, and PIT tags. Simple glue dots are very effective for tagging very small juveniles (Figure 7.4a). Black superglue works very well when the shell is clean, but not dry. Using a superglue accelerant to polymerize the glue is helpful. The primary drawback of simple glue dots is the inability to differentiate between

(a)

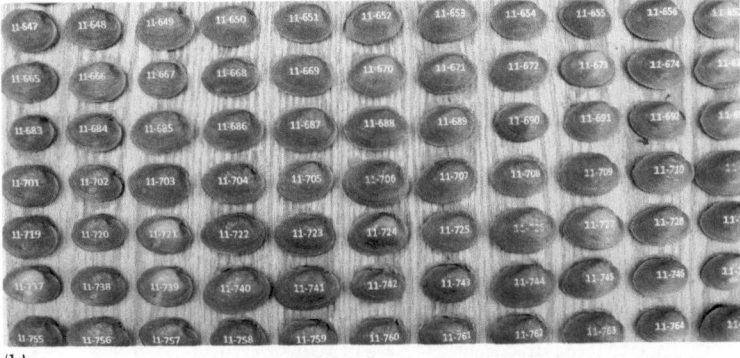

(b)

FIGURE 7.3 (a) A subadult alewife floater (*Anodonta implicata*) tagged
with a unique letter and number combination using a laser engraver.
(b) A large cohort of sub-adult pink mucket (*Lampsilis abrupta*) tagged
with a laser engraver. (c) Positioning the focal point of the laser on a
subadult alewife floater (*Anodonta implicata*) to ensure the etched tag
does not cut too deep and perforate the shell. (d) The laser engraver at
Virginia Fisheries and Aquatic Wildlife Center at Harrison Lake National
Fish Hatchery (Charles City, VA). A series of rubber washers are used to
create a grid inside the engraver to hold the mussels in position during the
engraving process. (e) The laser engraver at Missouri State University uses
a board with a grid of shallow depressions (18 × 12) to hold the mussels
in position during the engraving process. (f) A metal plate with shallow
depressions also can be used to hold mussels in place in the laser engraver.
Missouri State University noticed that the wooden board would warp
over time and replaced it with a metal plate. (g) The Epilog Zing-24 laser
engraver used at Missouri State University showing the grid, mussels, and
laptop used to set up the Excel spreadsheet with tag numbers.
Photos: (a, c, d) Matthew Patterson, USFWS. (b, e, f, g) Chris Barnhart,
Missouri State University. (A black-and-white version of part (b) of this
figure will appear in some formats. For the color version, please refer to
the plate section.)

(c)

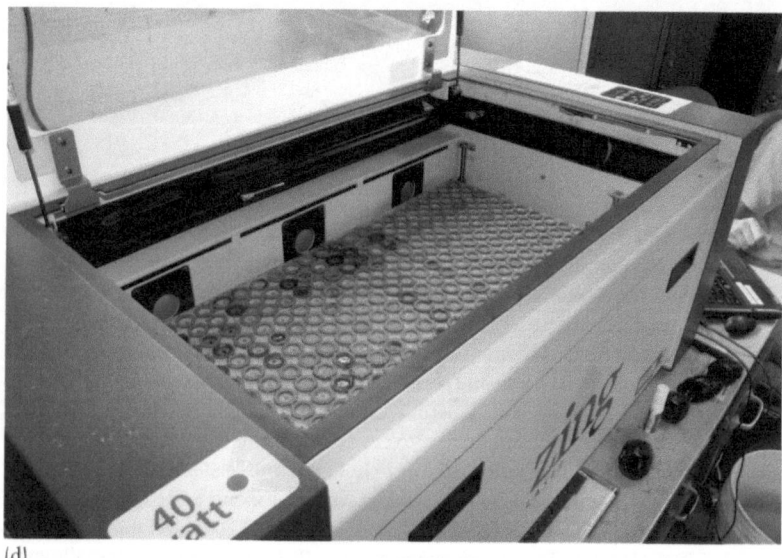

(d)

FIGURE 7.3 (cont.)

(e)

(f)

FIGURE 7.3 (cont.)

(g)

FIGURE 7.3 (cont.)

individual mussels. Hallprint glue-on shellfish tags are pre-printed, plastic tags that are superglued onto the shell (Figure 7.4b). The tags are available in different colors with unique letter-number combinations. With careful cleaning of the shell and proper gluing technique, they can stay on the shell for years. The West Virginia Division of Natural Resources attaches the same Hallprint tag (same color and number) to both the right and left valves in case the mussel dies and only one valve is recovered (Janet Clayton, WVDNR, personal communication). Hallprint tags are relatively expensive (about 25 cents each) but are widely used in mussel propagation. If different color Hallprint tags will be used to differentiate between cohorts of propagated juveniles, be aware that some colors may fade over time (Figure 7.4c). Field testing has shown that a secondary color like purple can fade to its primary color (blue) after being in the water for about 6 months (Jeremy Tiemann, INHS, personal communication).

(a)

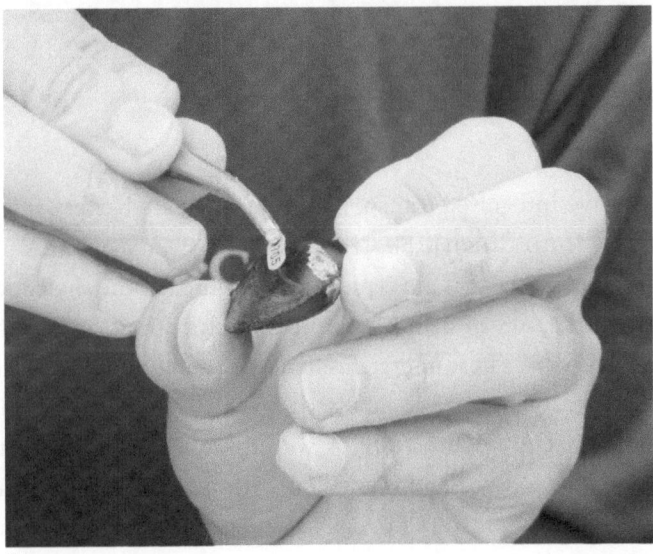

(b)

FIGURE 7.4 (a) Higgins eye pearly mussel (*Lampsilis higginsi*) tagged with both a black glue dot and a Hallprint tag. (b) A Hallprint tag being applied with superglue and forceps to the Eastern elliptio (*Elliptio complanata*). (c) A northern riffleshell (*Epioblasma torulosa rangiana*) tagged with a purple Hallprint tag that is starting to fade to blue in the river.
Photos: (a) Gary Wege, USFWS. (b) Matthew Patterson, USFWS.
(c) Jeremy Tiemann, Illinois Natural History Survey. (A black-and-white version of parts (a) and (c) of this figure will appear in some formats. For the color version, please refer to the plate section.)

(c)

FIGURE 7.4 (*cont.*)

Passive integrated transponder (PIT) tags are battery-free radio frequency identification (RFID) tags that are powered by transmission from the reader. PIT tags are typically superglued to the shell and then covered with epoxy or dental cement to protect the tags from break-age (Figure 7.5a). These inert glass encased microchips are encoded with unique alphanumeric labels that easily can be recorded with a handheld reader (Figure 7.5b). A waterproof PIT tag reader allows a mussel to be located and identified without removing it from the substrate (Figure 7.5c, d). PIT tags have recently gotten wider use in the United States, especially for imperiled species. Major drawbacks of PIT tags include high cost ($3–$5 per tag), long application time because of the two-step process of superglue and epoxy, and inability to identify the mussel without the reader. The inability to identify the mussel without a reader can be remedied by adding a second tag type like a Hallprint tag to the shell (Figure 7.5e).

Selection of the most appropriate tag will depend on the goal of the monitoring project and the species being tagged. If the monitoring protocol is designed to measure simple recapture efficiency or basic stocking success, a simple tag may be sufficient. A more complex mark–recapture study that utilizes individual data from individual

(a)

FIGURE 7.5 (a) PIT tags coated in white epoxy in combination with Hallprint tags can be used to monitor imperiled species like the northern riffleshell (*Epioblasma torulosa rangiana*). (b) Tagged mussels can be removed from the substrate and passed through the PIT tag reader on shore. (c) A PIT tag reader being used to detect tagged mussels from the surface in a small stream. (d) A PIT tag reader being operated by a SCUBA diver to detect tagged mussels in a large river. (e) An Eastern elliptio (*Elliptio complanata*) tagged with both a PIT tag and Hallprint tag. The Hallprint tag can be used to identify the mussel if the PIT tag reader is either not available or not functioning properly. Photos: (a) Rachel Mair, USFWS. (b) Sara Weglein, Maryland Department of Natural Resources. (c, e) Matthew Patterson, USFWS. (d) Janet Clayton, West Virginia Division of Natural Resources. (A black-and-white version of parts (a) and (d) of this figure will appear in some formats. For the color version, please refer to the plate section.)

(b)

(c)

FIGURE 7.5 (cont.)

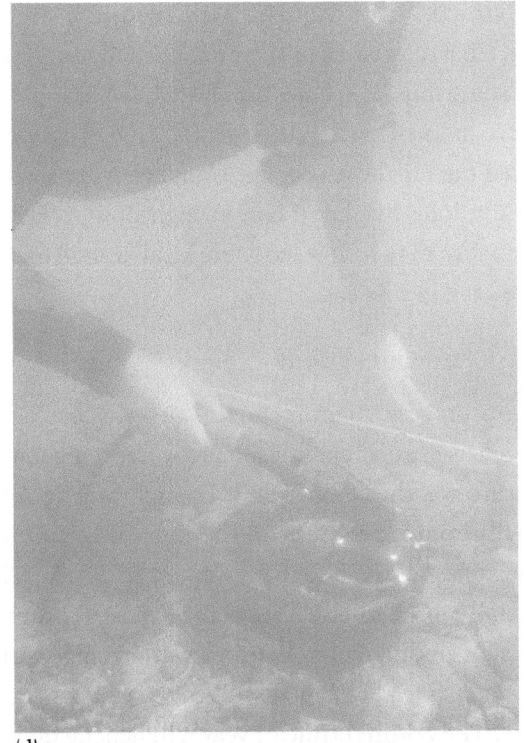

(d)

(e)

FIGURE 7.5 (*cont.*)

mussels (i.e. growth or movement) will require a more complex numbered tag. Characteristics of the mussel shell also will affect the choice of tag. Some mussel species are short-lived, fast-growing, thin-shelled, and live in low-flow environments. Glue or glue-on tags may be the preferred tag for these species. Conversely, thick-shelled species are typically long-lived and live in high-flow, volatile environments. Glue-on tags may be lost, so some kind of shell engraving may be the preferred tag in this case.

7.1.2   Release Site Conditions

Once the tagging process is complete, it is time to check current conditions at the release site, including air and water temperatures, water levels, and future weather forecasts. It is important to select a time of year that will minimize stress on the mussels. Summer air temperatures that exceed a daytime maximum of 35 °C (95 °F) should be avoided, especially if the mussels are being exposed to air for tagging, measurements, or photographs. Warmer water temperatures during the summer can also lead to low dissolved oxygen concentrations at the release site. The optimal time of year will depend on latitude and elevation of the release site, but early spring or fall, when air temperatures are cooler, seems to be the best time of year to minimize handling stress. Some practitioners have noted higher survival for mussels stocked in the fall compared to those stocked in the spring. Higher survival during fall releases could be the result of juvenile mussels having more time to grow to a larger size before release.

It is also important to check water levels prior to release. High-flow events typically carry heavy loads of suspended sediment that can clog the gills of juvenile mussels and interfere with feeding and respiration. Small juvenile mussels also could be entrained in the water column during high-flow events and carried downstream. Downstream movement could send mussels to an area of unsuitable habitat or out of the restoration and monitoring zone. Extremely low water conditions that can cause temperature spikes and periods of low dissolved oxygen also should be avoided. Find a reliable gauge

near the release site and keep a close eye on the weather before heading out to the river.

7.1.3 Biosecurity

If any pathogen at the propagation facility could potentially be spread
to the release site, all mussels should be held in clean, filtered water
for minimum of 4 hours to purge their guts prior to release. Gatenby
(2000) reported gut residence time in juvenile mussels between
2 hours and 6 hours depending on the feed ration. Many marine hatcheries purge the gut for two hours before sending shellfish to the market.

It is also very important to ensure invasive species are not
being spread during the release process. Aquatic invasive species (i.e.
zebra mussel, quagga mussel, New Zealand mud snail, rock snot,
apple snail, etc.) are causing ecological and economic damage all over
the country and steps should be taken to ensure mussel restoration
efforts do not contribute to the problem. Release is the final step in
the propagation process, so this is the opportune time to do a final
check for invasive species. Techniques that will help prevent the
spread of aquatic invasive species include pre-cleaning mussels with
Scotch-Brite pads, stiff bristle brushes or toothbrushes, visual inspections, and disinfecting mussel bags and other field gear.

7.1.4 Transport

If conditions are suitable, it is time to transport the mussels to the
release site. The size of the transport container needed will depend on
the size and number of mussels to be released. If a small number of
juvenile mussels (less than 100) will be transported, a simple cooler
with a portable aerator may suffice. If a large number of subadult
mussels are being hauled, a large container like a fish-hauling truck
may be needed (Figure 7.6a, b).

The water in the transport container should be monitored regularly during transport. At a minimum, water temperature and dissolved oxygen should be monitored. For long trips, ammonia and pH
also should be monitored. If water temperatures start to rise in the

(a)

(b)

FIGURE 7.6 (a) West Virginia Division of Natural Resources staff loading mussels into stackable plastic dish racks for transport in a fish-hauling truck. The racks help prevent mussels on the bottom of the hauling tank from being crushed. (b) Close-up of mussels loaded in the plastic dish racks.

Photos: Janet Clayton, West Virginia Division of Natural Resources.

(A black-and-white version of part (b) of this figure will appear in some formats. For the color version, please refer to the plate section.)

FIGURE 7.7 Ice can be added to a transport container to limit temperature fluctuations during transport. The ice should be double bagged to prevent residual chlorine from leaching into the transport water.
Photo: Matthew Patterson, USFWS.

transport container, non-chlorinated ice or gel ice packs can be added. If you are not certain the ice is non-chlorinated, double bag the ice in Ziploc containers to ensure any residual chlorine does not leach into the transport water (Figure 7.7). Aerators or liquid oxygen can be added if dissolved oxygen declines. Carrying extra source water also can be a good idea in the event a water change is needed.

Once at the release site, the transport water should be tempered to match the receiving water. Rapid shifts in temperature and pH can be very stressful for juvenile mussels. When handling imperiled species, permits may have specific requirements that limit exposure to extreme temperature shifts. After release, all transport water should be treated with 8% sodium hypochlorite and disposed of in an approved upland location.

7.1.5 Release Techniques

There is some debate in the freshwater mussel community about the preferred technique for placing adult and juvenile mussels into the substrate. Some biologists use a digging device to loosen the substrate before hand-placing the mussels in the river bottom. Others lay the mussels out on the substrate, allowing the mussels to find their preferred habitat. Loosening the substrate could help minimize energy expended on burrowing and limit exposure to predators by reducing time spent on the surface. Hand-placing juvenile mussels in poor habitat, however, could increase energy expenditures if they have to move to suitable habitat and re-burrow (or worse die in place). If mussels will be hand-placed in the substrate, it is critical to ensure they are placed in the correct orientation with anterior end in the substrate and the posterior end up toward the water surface.

Whatever release technique is employed, it is important to keep detailed records of all juvenile mussel releases. The basic information to record is included in the example Freshwater Mussel Propagation and Release Form and Trip Record (Figure 7.8a, b). The release section of the form includes fields for the date of release, specific details about the location of the release, tag numbers, water temperature, habitat description, and release method. There is also a field for additional notes that can be used to record any and all observations about the mussel release.

Mussel releases are also an excellent way to inform the public about the importance of freshwater mussels. If possible, invite the general public, partners, agency leaders, and local officials to a release event to increase awareness about freshwater mussels, and show partners and funding agencies how their money is being used to achieve tangible results on the ground.

FRESHWATER MUSSEL PROPAGATION & RELEASE FORM
UPPER TENNESSEE RIVER AUGMENTATION PROJECT

Organization: _____ Primary contact: _____

Address: _____ City: _____ State: _____ Zip: _____

Phone: _____ Email: _____

Adult Mussel Collection

Species: _____ Status: _____

Collection date: _____ Number propagated: _____

Hallprint tag ID and color: _____

Condition: _____

Collection method: _____ Drainage: _____

State: _____ County: _____ Quad: _____ River mile: _____

UTM X: _____ UTM Y: _____

Site location: _____

Habitat: _____

Collection crew: _____

Holding structure: _____ Holding duration: _____

Holding location: _____

Additional comments: _____

Fish Host Collection

Fish Host(s) Collected: _____

Collection date(s): _____ Drainage: _____ County: _____

State: _____ Quad: _____ River mile: _____

UTM X: _____ UTM Y: _____

Site location: _____

Habitat: _____

Collection method: _____

Fish disposal: _____

Collection crew: _____

Additional comments: _____

Mussel Release

Release date: _____ Augmentation reach: _____ Drainage: _____

County: _____ Quad: _____ River mile: _____

UTM X: _____ UTM Y: _____

Site location: _____

Habitat: _____

Number released: _____ Age at release: _____ Size range: _____

Mussel condition: _____

Release method: _____

Release crew: _____

Additional comments: _____

(a)

FIGURE 7.8 (a) Example freshwater mussel propagation and release form used by the Virginia Department of Game and Inland Fisheries. (b) Example trip record for the release of freshwater mussels into the wild.

Forms courtesy of: (a) The Virginia Department of Game and Inland Fisheries. (b) Genoa National Fish Hatchery, USFWS.

GENOA NATIONAL FISH HATCHERY
S 5631 State Hwy 35, Genoa, WI 54632
608-689-2605

TRIP RECORD

						TRIP NO.	DATE
SPECIES	LOT NUMBER	NUMBER	NO / LB	LENGTH	REARING UNIT	WEIGHT	# INCHES DISPLACED

STOCKING SITE

TEMPERATURE	DEPART	COMMENTS:
	DESTINATION	
	WATER BODY	
IF MORE THAN 10 F DIFFERENCE BETWEEN TANK AND WATER BODY, MUST TEMPER. LENGTH OF TIME:		
VEHICLE #		
DRIVER		

CERTIFICATION OF FISH STOCKING:

FISH AND WILDLIFE SERVICE REPRESENTATIVE: _____
(SIGNATURE)

WITNESS (OPTIONAL): _____
(SIGNATURE)

(b)

FIGURE 7.8 (cont.)

7.2 JUVENILE MONITORING

It will be difficult to evaluate the success or failure of a propagation program without long-term monitoring. It is important to define what success or failure will look like ahead of time by establishing clear goals and objectives for each species and each release site. Goals and objectives should be measurable, feasible both financially and scientifically, biologically meaningful, and compatible with the goals and objectives for other species. As a reminder, monitoring goals and objectives should be identified during the propagation planning process before propagation begins.

7.2.1 Scope, Duration, and Frequency of Monitoring

It may not be feasible to monitor all the release sites or even the entire extent of one release site for any given mussel species. So, it may be necessary to identify key areas of monitoring interest and then extrapolate the data to areas not being monitored. When extrapolating from a subsample, it is important that all samples are either systematically or randomly distributed across the monitoring area.

Monitoring of the release site should begin prior to release. Pre-release monitoring will provide baseline data on extant mussel populations at the release site and assist with the selection of quality release sites.

After the juveniles are released, there are differing opinions on the optimal frequency of monitoring. Monitoring too often can lead to undue stress on the mussel bed, while monitoring too infrequently could make it difficult to identify the causes of success or failure. Standardized guidelines for monitoring freshwater mussel restoration projects are not readily available; however, the oyster restoration community has identified a series of guidelines that may be helpful. *The Oyster Habitat Restoration Monitoring and Assessment Handbook* (Baggett *et al.*, 2014) recommends a schedule that includes monitoring prior to release, within three months post-release and 1–2 years post-release. The authors also recommend monitoring 4–6 years post-release, with additional measurements taken after events that could potentially impact survival (e.g. storm events). The West Virginia Division of Natural Resources monitors species richness and mussel density at their long-term monitoring sites every 3–5 years, depending on water conditions (Janet Clayton, personal communication). This monitoring frequency falls within the recommended frequency for long-term monitoring of oyster restoration projects, so this schedule is likely a good model for freshwater mussel restoration projects.

7.2.3 *Sampling Design*

The goals and objectives of the monitoring program will largely determine the appropriate sampling design. Clear and measureable objectives in the Propagation Plan will make it easier to design a sampling protocol. Sampling designs vary from simple exploratory studies to intense probability-based quantitative studies (Strayer and Smith, 2003). Some basic techniques for sampling freshwater mussels include reconnaissance, qualitative, semi-quantitative, quantitative, and total coverage. Total coverage involves sampling the entire site to obtain a complete census of the mussel population. This technique is typically only used in special situations (i.e. research or relocation) because it is very expensive and time-consuming. Reconnaissance, qualitative, and semi-quantitative sampling are all informal sampling designs. The drawback of informal sampling design methods is that the data collected cannot be extrapolated to the entire mussel population. Informal approaches, however, can be used to quickly and cheaply determine mussel distribution or find rare species.

Quantitative sampling can take significantly more time, effort, and funding, but the data can be statistically compared across time and space (Figure 7.9). It is critical that the sample size (i.e. number of quadrats to be sampled at a site or number of passes at a site in mark–recapture) be determined prior to sampling. Sample size will greatly influence precision (coefficient of variation) of the estimated mean density of mussels at the release site. A small sample size will decrease precision and make it difficult to detect changes in mussel density and survival in the future. Knowledge of expected survival and emigration rates of mussels from the release site will be needed to calculate the ideal sample size for monitoring.

In the end, it is likely that some combination of qualitative and quantitative methods will be the most efficient method of reaching the monitoring objectives. This book will not address the fine details of

(a)

(b)

FIGURE 7.9 (a) Quantitative sampling for freshwater mussels. The substrate is excavated from the PVC quadrat and sieved to search for adult and juvenile mussels. (b) Completing a timed sample of a long-term monitoring grid on the Cacapon River in West Virginia. The orange flags indicate the position of a freshwater mussel in the PVC grid. Photos: Matthew Patterson, USFWS. (A black-and-white version of this figure will appear in some formats. For the color version, please refer to the plate section.)

the various sampling design methodologies. When developing a monitoring plan for freshwater mussels, consult the literature and books like *A Guide to Sampling Freshwater Mussel Populations* (Strayer and Smith 2003). Several state agencies also have monitoring plans that can provide guidance on choosing an appropriate sampling design.

Baggett *et al.* (2014) recommend a series of universal metrics that should be sampled for every oyster restoration project, regardless of the restoration goal. A whole suite of other metrics can be measured, but the universal metrics, at a minimum, should be monitored for every restoration project. Consistent measurement of the universal metrics allows for the assessment of success or failure through time, while also providing a standard of comparison to other projects. The universal metrics they recommend for oyster restoration include reef area in m^2, reef height (m), oyster density (oysters/m^2), and oyster size–frequency distribution (percentage of oysters per size class). The authors also recommend measuring universal environmental variables for every restoration project. Environmental variables are used to assess impacts to the oyster reef and include water temperature, salinity, and dissolved oxygen. The measurement of reef height and salinity would not be useful metrics for freshwater mussel restoration projects, but the remaining metrics could be applied in freshwater mussel monitoring. All projects involving the propagation, release, and monitoring of freshwater mussels should at a minimum measure mussel bed area, mussel density, mussel size–frequency distribution, water temperature, and dissolved oxygen. It is also important to keep track of growth and survival of any tagged juveniles that have been released.

Any monitoring data that does not feed back into the propagation planning process is essentially wasted data. If the data is not used to help inform future decisions, then it is not clear why the data was collected in the first place. In the end, the data collected during monitoring should be used to assess if the goals and objectives in the Propagation Plan were met. If they were not met, then it is important to determine why they were not met and what steps need to be taken to improve results.

8 Building a Freshwater Mussel Propagation Facility

Catherine M. Gatenby and Nathan L. Eckert

Constructing and operating a freshwater mussel propagation facility can be a large and expensive undertaking; thus, careful thought must go into the design and development phase. Facility designs depend on a number of factors, including mussel species cultured, production goals, and local conditions, so this book will not offer a standard template to follow. With that said, this chapter will attempt to provide an introduction to the fundamentals of propagation facility construction and operations. Six over-arching elements will likely guide the planning and design process: (1) quality and quantity of the source water, (2) available infrastructure, (3) target species for propagation, (4) biosecurity, (5) space needs and the physical plant, and (6) economic considerations. This chapter will discuss all of these elements in more detail.

8.1 QUALITY AND QUANTITY OF THE SOURCE WATER

Water quality requirements for freshwater mussel physiology, growth, and reproduction are not well understood. The most important water quality parameters to consider in a potential water source are likely temperature, dissolved oxygen, adequate calcium for proper shell growth, and chemicals toxic to mussels.

Water temperatures over 20 °C will help maximize growth, but temperatures in excess of 34 °C in the summer months can cause mussel mortalities. Temperature can be controlled in the hatchery with chillers and heaters, but temperature manipulation comes with additional operational costs. While freshwater mussels are capable of surviving periods of low dissolved oxygen and even emersion, dissolved oxygen concentrations should be kept at 100% saturation to maximize growth and survival. Optimal calcium concentration in the water source will depend on the target mussel species. Hardness and alkalinity levels in the target species' native habitat can help establish optimal calcium concentrations for the propagation facility. Mussel propagation facilities with water hardness ranging from 100–350 ppm $CaCO_3$ have successfully reared freshwater mussels from the Ohio River, Delaware River, and James River drainages. The Virginia Fisheries and Aquatics Wildlife Center at Harrison Lake National Fish Hatchery successfully cultures juvenile mussels from the Atlantic Slope at hardness concentrations below 100 ppm $CaCO_3$ (Brian Watson, VDGIF, personal communication).

Freshwater mussels are sensitive to a variety of chemicals, especially during the juvenile life stage, that could be present in source water, including heavy metals, potassium, pesticides, and other contaminants (Naimo, 1995; Augspurger, *et al.*, 2003; Milam *et al.*, 2005; Newton and Bartsch, 2007; Wang *et al.*, 2007a, 2007b, 2010). Imlay (1973) showed that potassium concentrations of 11 ppm were lethal to *Actinonaias carinata*, *Lampsilis radiata siliquoidea*, and *Fusconaia flava* after 36–52 days of exposure while 7 ppm was lethal after 8 months. Potassium concentrations at the VFAWC (8–10 ppm) were found to be toxic to juvenile mussels after about 30 days of grow-out (Brian Watson, personal communication).

Potential negative impacts to the water source from land use practices in the watershed also should be evaluated. Upstream discharge events, high sediment loads, flooding, and periods of low dissolved oxygen could significantly affect the quality of the source water. If a new propagation facility has concerns about negative impacts to the source water, it may be necessary to drill a well on-site or select a different location for the facility.

A wild-water source (i.e. river, pond or reservoir) that supports an existing mussel population can help meet seasonal chemical and temperature requirements, as well as provide a wide variety of food resources. Wild water has the added benefit of reducing operational costs associated with purchasing food or producing food on-site.

Unfortunately, mussels may be exposed to any water quality issues occurring upstream of the facility as mentioned above.

Hatchery ponds can offer a stable water and food source for mussel culture that is secure from outside impacts (Figure 8.1). Water temperatures, however, may fluctuate seasonally (or daily) and may differ from seasonal temperatures in the wild, potentially leaving mussels exposed to temperatures outside their tolerance limits. Depending on the location of the facility, ponds also may freeze over in the winter. Pond temperature can be managed with the addition of well water at particular times of year. Fertilization may be necessary to increase the density of algae in the pond, but blooms of blue-green algae should be avoided. Adjusting the ratio of nitrogen to phosphorus can help limit the growth of blue-green algae. Freshwater mussels are very sensitive to nitrogen, and potassium is toxic to mussels above 11 ppm, so fertilizing a pond while mussels are present is not

FIGURE 8.1 Hatchery ponds at the Virginia Fisheries and Aquatic Wildlife Center at Harrison Lake National Fish Hatchery used to culture juvenile freshwater mussels in floating baskets.
Photo: Matthew Patterson, USFWS.

recommended. If fertilization is needed, a fertilizer that does not include potassium should be used.

Mussels also can be cultured in well water with the appropriate water chemistry. The mussel propagation facility at White Sulphur Springs National Fish Hatchery (White Sulphur Springs, WV) has been culturing freshwater mussels in well water for over 10 years. Well water from a cold-water facility like White Sulphur Springs National Fish Hatchery may require warming to be suitable for mussel culture. By contrast, a warm-water facility may need to chill incoming water in warmer months. The costs of heating and cooling water can be significant, but should not eliminate a facility from consideration. In mussel culture, water is often recirculated to keep algal food resources in suspension and avoid dumping precious food down the drain. In recirculating systems, cold well water will slowly increase to ambient air temperatures, aided by the heat released from the recirculating pumps. Well water also is limited in natural food resources, so algae will need to be cultured on station or purchased.

The quantity of water available from some sources can fluctuate seasonally. When selecting a site for mussel propagation, be sure the water needs of the facility will be met, even during low-flow events. Site selection also should include a plan for future water demands as the program grows. If the facility has a continuous wild-water source or has significant pond-water resources, then water quantity may not be a concern. Wild-water sources, however, can be vulnerable to low dissolved oxygen, chemical spills, and drought. One or more large wells may be needed to supplement during such events. If freshwater mussels are cultured in recirculating systems, the quantity of water needed can be significantly reduced.

8.2 AVAILABLE INFRASTRUCTURE

Existing fish hatcheries can provide excellent opportunities to build a mussel propagation program without constructing all new facilities (Figure 8.2). In many cases, the location for a hatchery was chosen based on the presence of a high-quality water source. Many hatcheries

(a)

(b)

FIGURE 8.2 (a) The exterior of the administrative and laboratory building prior to renovations at the Alabama Department of Conservation and Natural Resources' Alabama Aquatic Biodiversity Center in Marion, AL. (b) The exterior of the administrative and laboratory building after renovations. (c) The interior of the wet lab and culture area prior to renovations. (d) The interior of the wet lab and culture area after renovations.

Photos: Paul Johnson, Alabama Aquatic Biodiversity Center.

(c)

(d)

FIGURE 8.2 (cont.)

have extra space and equipment for holding mussels and host fish, and even may be able to produce the host fish needed for mussel propagation. Existing maintenance staff skilled in plumbing and other hatchery operations can be invaluable to a mussel propagation program. The quality of the water source, available space, and biosecurity issues, however, should be carefully evaluated when considering an existing facility. If space is limited or the water source is suspect, substantial time and funding may be needed to remedy the situation, which could delay propagation efforts. Several national and state fish hatcheries in the United States have started freshwater mussel propagation programs. Many propagation programs constructed new buildings on the hatchery property, while the existing water, utilities, and maintenance staff were shared with existing programs.

8.3 TARGET SPECIES FOR PROPAGATION

The mussel species targeted for propagation can have an impact on site selection. Propagation efforts are typically more efficient and successful when the propagation facility is located close to wild populations of the target species. Traveling long distances to collect gravid female mussels will increase costs and reduce the amount of time staff spend propagating mussels in the laboratory. Riusech and Barnhart (2000) also showed higher rates of metamorphosis for glochidia attached to fish hosts collected from the same drainage as the gravid females. Consequently, building a facility close to the target species also could save travel time for collection of suitable host species. Long-distance transport of the gravid mussels and host species also can increase stress levels and reduce survival in the laboratory.

8.4 BIOSECURITY

As discussed in Chapter 3, biosecurity is a major concern when bringing wild fish and potentially wild pathogens onto an existing facility. When building the freshwater mussel propagation facility at the White Sulphur Springs National Fish Hatchery (White Sulphur Springs, WV), all potential impacts to the hatchery's Class A fish health certification were carefully considered. Maintaining the hatchery's status as disease-free was critical to the station mission and the mission of the US Fish and Wildlife Service's national brood stock program. The mussel propagation facility was built as far away from the fish culture buildings as possible and equipped with a separate vehicle entrance. All vehicles used for the mussel propagation program enter and exit the hatchery from a rear entrance, never driving through the fish culture facility. Disease-free egg shipments and fish-hauling trucks enter and exit from a front entrance. If transportation and entrance/exit strategies had not been considered during the design phase, a more time-consuming decontamination protocol would have been required to prevent the spread of disease. A dedicated quarantine space for all wild fish arriving on station also was built to further limit the spread of pathogens. Finally, all flooring and wall surfaces of the mussel propagation facility were covered with a durable, mildew-resistant finish to facilitate cleaning.

If a mussel propagation facility is located along the banks of a river, the most logical design would be to use river water for culturing mussels. Unfortunately, pumping wild river water could expose the facility to invasive species or pathogens. Exposing propagated mussels to invasive species and novel pathogens also could limit where those mussels are stocked in the future.

8.5 SPACE NEEDS AND THE PHYSICAL PLANT

The amount of space needed will depend on the restoration goals set out in the Propagation Plan (i.e. target species, the number of each species to be released, and the size at release), culture methods (indoor or outdoor culture and the duration of culture) and the type of culture system (downwellers, upwellers, cages, baskets, pans, etc.). For example, juvenile mussels that will be cultured to a large enough size to tag prior to release will have to spend more time in the facility, increasing space needs. If ponds will be used for juvenile mussel grow-out, additional acreage with suitable soils will be needed. Refer to previous chapters for detailed information on the various systems used for juvenile mussel culture, and the maintenance of brood stock and host species.

If large numbers of mussels of multiple species are being cultured to taggable size, multiple buildings that support both large numbers of host fish and multiple life stages of mussels may be needed. If mussels will be collected from multiple drainages, extra space will be required to maintain culture systems with discrete water filtration, water treatment and discharge to prevent potential spread of disease inside the facility (as well as to prevent mixing of stocks of juvenile mussels). Dedicated space for wild fish quarantine also will be needed. The White Sulphur Springs National Fish Hatchery constructed two, 3000 square foot buildings, one dedicated for host fish and the other for mussel culture (Figure 8.3). If a relatively small number of juvenile mussels from a single species will be produced, then a

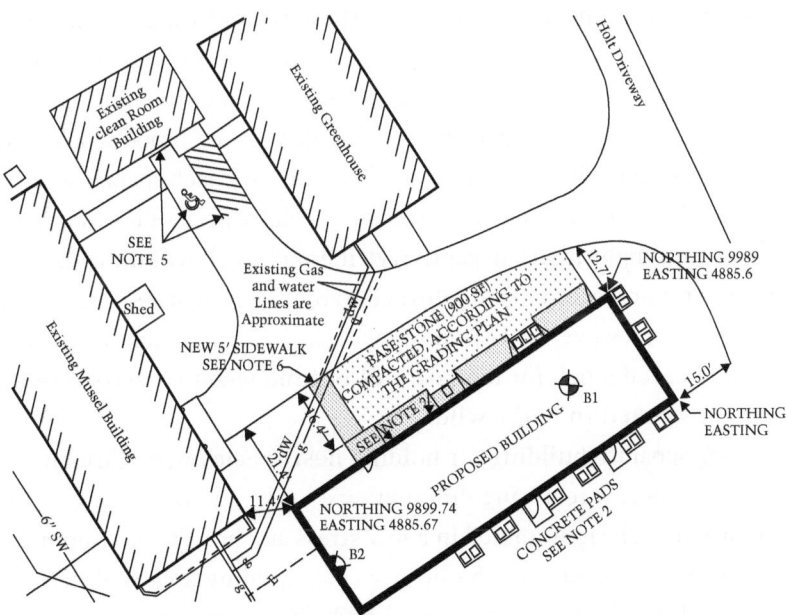

FIGURE 8.3 Engineering schematic of the mussel culture facility at White Sulphur Springs National Fish Hatchery showing the mussel building, wild fish building, greenhouse and clean room. Schematic provided courtesy of the USFWS.

small culture building or laboratory may be sufficient. If infested fish will be released into cages or free into the river, space needs will be further reduced.

When considering space needs, it is important to plan for future needs and flexibility so the hatchery can adapt to changes without major rebuilding. For example, plastic or fiberglass tanks are preferred to concrete as they can be easily moved or changed as needed. Floors should be constructed of concrete and have sufficient drains. Considering the costs associated with design and construction, it may be more economical in the long run to construct a larger building than needed to meet current propagation goals.

The size of the brood stock holding area will depend on the number of mussels held in captivity from different river drainages at any given time, the number of different mussel species held in captivity at any given time, the total number of brood stock held in captivity at any given time, and the length of time brood stock will be held in captivity. It is important to keep mussels from different drainages in separate aquaculture systems to help prevent the spread of invasive species and pathogens when brood stocks are returned to the wild. Individual holding tanks for each species within a given drainage also may be needed (especially for species that tend to release conglutinates in captivity). If brood stock will immediately be used for propagation and returned to the river, then less space will be needed. However, it is important to have a contingency plan for holding brood stock for long periods of time when river conditions prevent safe return to the wild.

A separate building for holding host species will allow many different species requiring different size tanks and different culture conditions to be maintained in a low stress and healthy environment (Figure 8.4). The size of the host species holding area will depend on how long the host species will be held in captivity, the species being held in captivity, and the number of hosts being held in captivity. If a host species will be held for a short period prior to infestation and released immediately after juvenile drop-off, then less space

FIGURE 8.4 Construction of the wild fish building at the White Sulphur Springs National Fish Hatchery. This 3000 square foot building was designed to increase juvenile mussel production by increasing the space available for holding host fish.
Photo: Matthew Patterson, USFWS.

may be required. If a host will be held long term with temperature manipulation to extend the larval attachment period, more space will be needed. Different host species also require different size holding tanks. Walleye, bass, and freshwater drum, for example, require large tanks and thus more space than minnows and darters. Long-term holding of host species also may require larger tanks with more rearing space to minimize the effects of long-term stress. If all available space is occupied, new host-fish infestations will be delayed. Eventually, mussel propagation programs that infest host fish will reach a plateau of juvenile production based on available space for holding host fish.

The amount of space dedicated to juvenile culture varies widely among existing mussel propagation facilities (Figure 8.5). Research facilities typically commit very little space to juvenile culture, while large-scale production facilities might dedicate a significant percentage of the available square footage. An intensive culture facility may require 25% or more of the floor plan for juvenile culture, whereas a free release or cage culture facility will dedicate little space for juvenile culture. The floor plan should be designed to accommodate all projected propagation activities, but allow some flexibility to expand and test new technologies as they become available.

(a)

FIGURE 8.5 (a) Construction of the Virginia Fisheries and Aquatic Wildlife Center's mussel culture building at Harrison Lake National Fish Hatchery. (b) The exterior of the 480 square foot mussel culture building at the Virginia Fisheries and Aquatic Wildlife Center. A smaller culture building was sufficient at this facility because the Center uses outdoor ponds for juvenile grow-out. (c) The interior of the mussel culture building at the Virginia Fisheries and Aquatic Wildlife Center.
Photos: Michael Odom, USFWS.

(b)

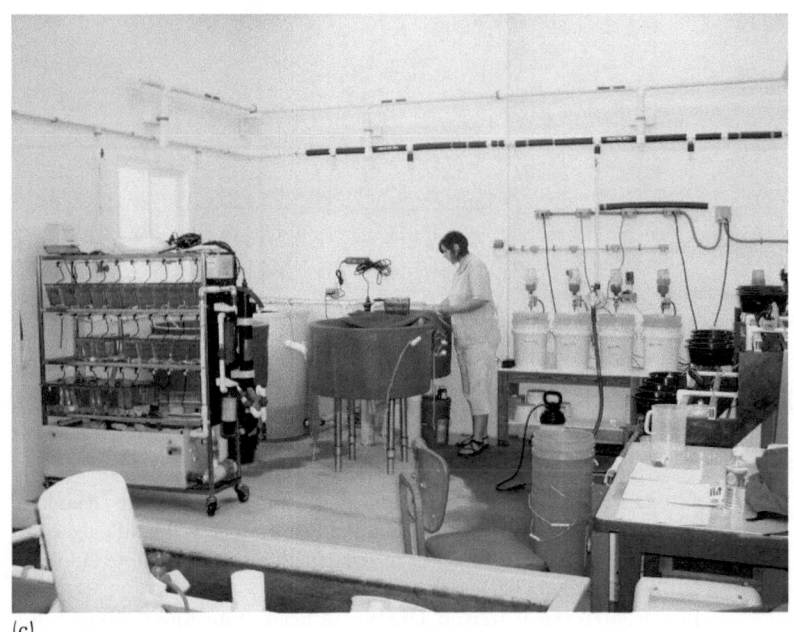

(c)

FIGURE 8.5 (*cont.*)

If algae will be cultured on-site, space will have to be set aside for a clean area to maintain axenic stock cultures and a greenhouse or other facility for growing bulk cultures (Figure 8.6). An algae culture facility can prove to be valuable for intensive culture of juveniles and for long-term holding of adult mussels, especially when the water

(a)

FIGURE 8.6 (a) Algae culture slants, flasks, and carboys in the clean lab at the White Sulphur Springs National Fish Hatchery. (b) Bulk algae cultures growing in a Biofence in the greenhouse at the White Sulphur Springs National Fish Hatchery.
Photos: Matthew Patterson, USFWS. (A black-and-white version of this figure will appear in some formats. For the color version, please refer to the plate section.)

(b)

FIGURE 8.6 (*cont.*)

source is well water or oligotrophic surface water. If biosecurity concerns prevent the pumping of wild water onto the facility, algae culture may be the only option. Labor costs should be considered when deciding to build an algae culture facility. Preparing, culturing, and maintaining algae cultures can be very time-consuming. Once the algae culture facility is constructed and up and running, however, live algae cultures can be a very nutritious and relatively inexpensive food source. If strict protocols for algae culture maintenance are not followed, the cultures will likely crash, greatly increasing time and cost.

Constructing a new pond on the facility will require a significant amount of space, but can provide numerous benefits. In addition to juvenile mussel culture, ponds can be used to warm well water, propagate fish, and hold host fish for mussel production. At most latitudes, pond temperatures will stay between 20 °C and 30 °C for most of the growing season. In areas of extreme heat or cold, special attention will have to be given to water temperatures to achieve optimal juvenile growth. The pH cycle and pond design (lined vs. unlined, gravel vs. mud) also can play a role in temperature and productivity. The space needed for pond culture will depend on the size of the pond. There is debate about the appropriate size of a pond for mussel culture. No data exists regarding optimal size, but freshwater mussels have been successfully reared in ponds ranging from 0.25 up to 5 acres.

8.6　ECONOMIC CONSIDERATIONS

Even if an existing facility needs extensive renovation, it may be more economical than constructing new buildings. Minor renovations like widening doorways, pouring small concrete slabs, or constructing aquaculture systems oftentimes can be completed by hatchery staff at significant time and cost savings relative to hiring a contractor. The Virginia Department of Game and Inland Fisheries renovated two abandoned fish culture buildings to construct the Aquatic Wildlife Conservation Center in Marion, Virginia for around $100 000.

Operational costs at an existing facility can be used to estimate any additional operational costs associated with a new mussel propagation program. Small-scale mussel propagation programs will likely add minimal electrical and heating costs. Mussel propagation, however, can be very labor intensive, with both fieldwork to collect host fish and brood stock from the wild, and laboratory work to maintain brood stock and host fish health in captivity. Consequently, adding a mussel propagation program could substantially increase staffing costs. For large, production-scale mussel propagation programs, at least three full-time staff are needed, especially during the field season. Four full-time staff would be ideal with one person for host species maintenance, one person to monitor the juvenile mussels and adult brood stock, and two persons to conduct fieldwork.

Constructing new buildings may be more expensive and time-consuming than retrofitting existing hatchery buildings, but the facility can be designed to meet long-term mussel production goals. One of the early mussel propagation facilities built in the United States in the early 2000s, cost upwards of 1.5 million dollars. The facility included a high-capacity back-up generator, well, ponds, clean lab, greenhouse, host-fish building, and mussel building for juvenile culture and brood stock maintenance, as well as all mechanical, electrical, water filtration, and culture systems for fish, mussels, and algae. Construction costs for this facility were incurred over a 5 year period. Since that time, multiple agencies have constructed mussel propagation facilities at existing hatcheries for between $100 000 and $400 000, including all utilities and well construction.

If the goal is to produce a small number of mussels from only a few species, a streamside facility may be ideal. Streamside facilities typically pump water from a stream that supports a healthy mussel assemblage so the costs associated with maintaining water quality and providing food are typically negligible. Costs for building a streamside facility depend upon the size of the unit, but a typical unit can be built for approximately $100 000. Once a streamside facility is completed, staff time and electricity are the only operational costs. Depending on the level of care required, staffing needs can range from dedicating one full-time employee to the facility down to short, daily visits. When selecting a location for a streamside facility, the unit should be close enough to the water source for a pump to function properly while resting on high enough ground to avoid potential flood waters. Mobile streamside facilities (i.e. trailers) provide the flexibility of moving prior to flooding, as well as moving to other drainages.

APPENDIX A
Genoa National Fish Hatchery
Fish Infestation Monitoring Form

Unit: _____ Tank: _____ Mussel species: _____ Tag ID: _____
Parental Source Location River: _____ Tributary: _____ Site: _____
Fish host: _____ Locality: _____ Number: _____ Infestation date: _____
Size: ____ Infestation Start: _____ End: ____ Total: _____minutes Researcher(s): _____

Date	Fish (live/dead)	Juveniles produced*	Running Total	Notes

nc=not checked

ID: _____ Location: _____ Subloc.: _____ Total to date: _____ Grand total: _____
Comments:
_____
_____

Batch ID: _____ Location: _____ Subloc.: _____ Total to date: _____ Grand total: _____
Comments:
_____
_____

Batch ID: _____ Location: _____ Subloc.: _____ Total to date: _____ Grand total: _____
Comments: _____
_____

Batch ID: _____ Location: _____ Subloc.: _____ Total to date: _____ Grand total: _____
Comments: _____
_____

APPENDIX B

Genoa National Fish Hatchery Juvenile Mussel Monitoring Form

Location ID:_____ Sublocation ID:_____ Mussel species: _____ Batch ID:____

Parental Source Location   River:_____ Tributary:_____ Site: _____

Project: _____ Date Produced: _____ Parent Mussel:_____

Date	Mussels Live	Mussels Dead	% Survival	Mean Length	Mean Height	MMW #	MMW Page#	Notes (Health, activity, predators,etc.)	Initials

Date Moved/Split: _____ Number Moved: _____ Age: _____ Yr ____ Mon _____ Day
New Batch ID: _____ New Location ID: _____   New Sublocation ID:_____
Comments:_____

Date Moved/Split: _____ Number Moved: _____ Age: ____Yr _____ Mon ____ Day
New Batch ID: _____ New Location ID: _____ New Sublocation ID:_____
Comments:_____

Date Moved/Split: _____ Number Moved: _____Age: ____Yr _____ Mon ____ Day
New Batch ID: _____ New Location ID: _____ New Sublocation ID:_____
Comments:_____

APPENDIX C
Genoa National Fish Hatchery Datasheet for Brood Stock Check-In

River & Locality: Date:

Collection Method:				Collectors:			Transport Method:			
Stream Conditions:						Project:				
						Disposition				
Species	Tag ID	Tank	Condition	Age	Sex	L	W	H	Gravid	Used	Removed
Comments:											

APPENDIX D
Freshwater Mussel Propagation Facilities in the United States

Facility	Agency	Contact Person	Contact Phone	Contact Email
Alabama Aquatic Biodiversity Center	ADCNR	Dr. Paul Johnson	334-683-5000	Paul.Johnson@dcnr.alabama.gov
Aquatic Epidemiology and Conservation Lab	NCSU	Chris Eads	919-270-3987	Chris_Eads@ncsu.edu
Aquatic Mollusk Programs	MNDNR	Bernard Sietman	651-314-6305	Bernard.Sietman@state.mn.us
Aquatic Wildlife Conservation Center	VDGIF	Tim Lane	540-354-5154	tim.lane@dgif.virginia.gov
Center for Mollusk Conservation	KDFWR	Dr. Monte McGregor	502-221-1379	monte.mcgregor@ky.gov
Cumberland River Aquatic Center	TWRA	David Sims	615-293-5237	David.M.Sims@tn.gov

Facility	Agency	Contact Person	Contact Phone	Contact Email
Freshwater Mollusk Conservation Center	VT	Jess Jones	540-231-2266	jess_jones@fws.gov
Freshwater Mussel Conservation & Res. Center	CZA	Dr. G. Thomas Watters	614-292-6170	Watters.1@osu.edu
Freshwater Mussel Rearing Facility	Mill of Kalborn	Dr. Frankie Thielen	35226908127	f.thielen@luxnatur.lu
Genoa National Fish Hatchery	USFWS	Nathan Eckert	608-689-2605 x 115	nathan_eckert@fws.gov
Harrison Lake National Fish Hatchery[a]	USFWS	Rachel Mair	804-829-2421 x 151	rachel_mair@fws.gov
Institute for Great Lakes Research	CMU	Dr. Dave Zanatta	989-774-7829	zanat1d@cmich.edu
Marion Conservation Aquaculture Center	NCWRC	Rachel Hoch	828-659-3324	rachel.hoch@ncwildlife.org
Missouri State University	MSU	Dr. Chris Barnhart	417-836-5166	chrisbarnhart@missouristate.edu

Facility	Agency	Contact Person	Contact Phone	Contact Email
Southeast Ecological Science Center	USGS	Nathan Johnson	352-264-3574	najohnson@ usgs.gov
Virginia Fisheries and Aquatic Wildlife Center[a]	VDGIF	Brian Watson	434-941-5990	Brian.Watson@ dgif.virginia.gov
Welaka National Fish Hatchery	USFWS	Tony Brady	386-467-2374	tony_brady@ fws.gov
White Sulphur Springs National Fish Hatchery	USFWS	Tyler Hern	304-536-1361	tyler_hern@fws. gov

[a]State and Federal Cooperative Unit located at Harrison Lake National Hatchery.

APPENDIX E
Algae Culture for Freshwater Mussel Propagation: A Companion Manual to the Online Video Tutorial

The production and delivery of a nutritious food source is a critical step in freshwater mussel propagation. The marine bivalve industry has tried many food types over the years, but no suitable food resource has been found that replaces live phytoplankton. The following pages are not intended to be a stand-alone guide to algae culture, but are designed to be a companion manual to the online video tutorial *Algae Culture for Freshwater Mussel Propagation*.

This manual describes the algal culture methodology designed by Drs. David and Pam Orcutt under contract with the White Sulphur Springs National Fish Hatchery (US Fish and Wildlife Service). The method uses small stock cultures to inoculate successively larger cultures until a sufficient amount of algae is cultured to supply the hatcheries needs.

The manual has a series of instructional videos that can be found at the link below.

http://nctc.fws.gov/resources/knowledge-resources/video-gallery/algae-culture.html

E.I STERILE TECHNIQUE STANDARD METHODS

It is very important when culturing algae for freshwater mussel propagation to work with pure cultures (pure starter cultures can be obtained from a variety of commercial sources). Unfortunately, the world around us is covered with microorganisms. Algae cells and bacteria cells are even carried on dust particles in the air. In order to protect sterile media, plates, slants, flasks, and carboys from these microorganisms, it is important to practice sterile (aseptic) technique at all times. This simply means that sterile surfaces or sterile media must be protected from contamination by microbes in the air or residing on non-sterile surfaces. In sterile technique, only sterile surfaces touch other sterile surfaces and exposure to the air is kept to an absolute minimum.

To maintain sterile cultures always do the following:

(1) Thoroughly clean and autoclave all glassware
(2) Autoclave or filter sterilize all media
(3) Wipe down all work surfaces with alcohol
(4) Use sterile pipettes and loops for transferring cultures into fresh media
(5) Do not touch sterile surfaces with your hands or any other non-sterile utensil
(6) When possible, always use the laminar flow hood when transferring cultures.

The laboratory where culture media is prepared and algae cultures are transferred must be kept as clean as possible. Counter tops, the inside of the laminar flow hood, floors, etc. should be cleaned and vacuumed at least weekly, if not more frequently, and the door to the laboratory should be kept closed at all times.

E.2 GETTING STARTED

E.2.1 Glassware Cleaning

All glassware used in algae culture must be very clean. Wash all glassware with a detergent that is designed to remove grease. Using the appropriate size brush, scrub all surfaces inside and out. If a piece of glassware is particularly dirty or has algae that has dried to the surface, it is best to let it soak in soapy water for an hour or more. After cleaning, rinse each piece of glassware in tap water at least 4 times inside and out. A final rinse should be done with distilled water. Invert the clean glassware on a wire rack to dry, making sure it is in a position where all water will drain out and it will dry completely.

After cleaning and drying, glassware should be sterilized in an autoclave. Glassware to be sterilized in the autoclave should be covered with a piece of heavy duty aluminum foil large enough to cover down the outside of the glassware at least one inch. Place a small piece of autoclave tape on each foil cover and mark the day, month and year using a black marker.

Bottles should be capped loosely. A piece of dated autoclave tape should be placed on the cap reaching to the bottle neck. Test tubes to be used for algae culture slants should be handled in the same manner.

Flasks for growing stock cultures (500 mL and larger) need to have a foam stopper, appropriately sized for the flask, inserted and covered with a large piece of foil in such a way as to not compact the foam stopper. Again a piece of dated autoclave tape is put on the foil top.

Bubbling tubes, feeding tubes, forceps, and other assorted items should be placed in the appropriate sized autoclavable bag and dated.

E.2.2   Sterile Distilled Water

Sterile distilled water should be used for algae culture. To make sterile distilled water, simply add distilled water to 1 L media bottles, leave the caps loose, and place a piece of dated autoclave tape on the cap. Autoclave and then allow to cool completely before tightening the lids.

E.2.3   Autoclave

Please follow your manufacturer specifications for sterilizing glassware and liquid media in an autoclave. The extremely high temperatures and pressures in an autoclave can be very dangerous if not operated properly.

E.2.4   Laminar Flow Hood

A laminar flow hood is important for any algae culture lab. As with the autoclave, please follow your manufacturer specifications for operation and maintenance of the laminar flow hood.

Follow the instructions below to prepare the laminar flow hood before working with algae cultures.

(1) Clean the inside of the laminar flow hood thoroughly with 70% isopropanol
(2) Once clean, turn on the UV light and blower and allow to sterilize for at least 15 minutes prior to use
(3) Once sterile, leave the blower on and turn off the UV before working under the hood.

E.3   MEDIA PREPARATION

There are a large number of algae growth media formulas available from commercial vendors. Below is the growth media formula used at White Sulphur Springs National Fish Hatchery.

E.3.1   F/2 Modified Guillard's Medium

Part A – contains trace nutrients and vitamins. This must always be filter sterilized.
Part B – contains primary nutrients but no silicate.
Part C – contains silicate.

For concentrated F/2:

(1) Using a 1 gallon glass jug and 1 L graduated cylinder measure out 3 L distilled H_2O into glass jug (jug and graduated cylinder do not need to be sterile). Place a piece of tape on jug and mark 3 L level. Remove about 1 L H_2O from jug by pouring into a 1 L flask or graduated cylinder.
(2) Add a stir bar and place jug on stirrer/hotplate with heat on 3.
(3) With funnel in the top of the jug slowly add the contents of one pack of Part B. It will take a while to get all added. Make sure what you have added goes into solution before adding more. Rinse bag with a little distilled H_2O and add to jug.
(4) Remove stir bar and add distilled H_2O to bring the total volume in jug to your 3 L mark.
(5) Using the 1 L graduated cylinder, dispense 980 mL of F/2 into each of two 1 L bottles. Then dispense the remaining L into two 500 mL

bottles, adding approximately 490 mL to each. If there is extra F/2 divide it between the two 1 L bottles. (It is better to store in smaller volumes. This decreases the times each bottle is opened and lessens the chance of contamination).

(6) Label bottles with contents and date, and place a strip of autoclave tape over corner of label. Add strip of autoclave tape over cap and bottle.

(7) Autoclave bottles on slow exhaust and wait until autoclave has cooled to below 100 °F to open door (helps prevent any boiling over and loss of media).

(8) Allow bottles to cool enough to be held in your hand before adding vitamins (Part A). Bottles can also be refrigerated as is and the vitamins added later. Do not tighten caps completely until totally cool or cap liners may be pulled out.

(9) Measure total volume of Part A, (this should be about 6 mL). Under the laminar flow hood, filter sterilize all of Part A into sterile capped test tubes and then using aseptic technique add one-third of total Part A (should be about 21.3 mL) to each of the two sterile 1 L bottles of F/2 and one-sixth of total Part A (should be about 10.6 mL) to each of the two 500 mL bottles of F/2.

(10) Add the words plus (+) vitamins to the bottle label.

(11) Cap tightly and refrigerate bottles in the dark (helps prevent degradation of vitamins).

(12) Use at the rate of 1 mL/L to make up final solution for carboys, flasks, or carboy tubes.

For F/2 with silicate:

(1) Add concentrated silicate at the rate of 0.66 mL/L to make carboys, flasks, or carboy tubes. Use aseptic technique in hood. Silicate will precipitate so shake well before use.

E.3.2 Essential Nutrients for F/2 Modified Guillard's Medium

Dr. Orcutt found commercially available F/2 Modified Guillard's Medium to be lacking in a few key ingredients or essential nutrients. The recipes for these essential nutrients are given below.

Essential Nutrients 1 (EN1)

In 1 L bottle	In 3 L glass jug
33.6 g H_3BO_3	100.8 g H_3BO_3
50.0 g $MgSO_4$ $7H_2O$	150.0 g $MgSO_4 7H_2O$
5.0 g KH_2PO_4 $7H_2O$	15.0 g KH_2PO_4 $7H_2O$
175.0 g $NaNO_3$	525.0 g $NaNO_3$

(1) Add approximately 800 mL distilled H_2O to a 1 L bottle or 2500 mL to a 3 L glass jug.

(2) Add stir bar and place on stirrer/hot plate. Set heat on 3–4 and stir.

(3) Weigh out each ingredient and add to the bottle in the order listed above. Make sure each is dissolved before the next ingredient is added. Add each slowly so they dissolve completely.

(4) Remove stir bar and using distilled H_2O bring volume in bottle to 1 L using calibrations on bottle.

(5) Label with contents and date. Place strip of autoclave tape over one corner of label. Cap loosely, place strip of autoclave tape over lid to bottle and autoclave on slow exhaust.

(6) Wait until autoclave temperature is 100 °F before opening door to prevent boiling over.

(7) Cool completely, then tighten cap and store. Does not need to be refrigerated.

USE AT THE RATE OF 1 mL/L IN FINAL MEDIA SOLUTION

Essential Nutrients 2 (EN2)

In 0.5 L bottle	In 3 L glass jug
25.0 g $CaCl_2$	150.0 g $CaCl_2$

(1) Add approximately 400 mL distilled H_2O to a 500 mL bottle or 2500 mL to a 3 L glass jug.

(2) Add stir bar and place on stirrer.

(3) Weigh $CaCl_2$ and slowly add to bottle while stirring. Don't add all at once, but a little at a time until all is in solution.

(4) Remove stir bar and using distilled H_2O bring volume in bottle to 500 mL (or 3 L) using calibrations on bottle.

(5) Label with contents and date. Place strip of autoclave tape over one corner of label. Cap loosely, place strip of autoclave tape over lid to bottle and autoclave on slow exhaust.

(6) Wait until autoclave temperature is 100 °F before opening door.

(7) Cool completely, then tighten cap and store. Does not need to be refrigerated.

USE AT THE RATE OF 0.5 mL/L IN FINAL MEDIA SOLUTION

NOTE: EN1 and EN2 cannot be made in one bottle. Adding $CaCl_2$ to Essential Nutrients 1 causes a precipitate that will not go back into solution.

E.3.3   For Concentrated F/2 silicate

(1) In a 1 gallon jug with the 3 L volume marked, as is done for F/2 preparation, bring volume up to 3 L, mark this level if not already marked and then pour off about 500 mL of water.

(2) Add stir bar. Add the packet of silicate gradually and mix until dissolved. Add heat if needed.

(3) Remove stir bar and bring total volume in jug to 3 L mark. Mix well.

(4) Pour into two 1 L bottles and two 500 mL bottles.

(5) Label bottles with contents and date (Silicate conc. date). Place a strip of autoclave tape over corner of label.

(6) Cap loosely, add autoclave tape to cap and bottle, and autoclave on slow exhaust.

(7) Cool completely, tighten cap, and store. Does not need to be refrigerated.

(8) Use at the rate of 0.66 mL/L to make up carboys or flasks using sterile technique in hood. Silicate will precipitate so shake well before using.

E.4   ALGAE SLANT CULTURE

E.4.1   Back-up Schedule

Healthy stock algae slant cultures should be transferred to fresh media every 2 weeks. If cultures do not look good, they may need to be transferred earlier.

Back-up protocol (see video):

(1) Using a sterile loop, transfer algae from the stock culture to a fresh, sterile agar slant.
(2) Streak the algae back and forth across the slant from top to bottom.
(3) Using a new sterile loop repeat the steps above for all species in culture.
(4) Leave the cap loose to allow for gas exchange.
(5) Place fresh cultures back in incubator.

E.4.2   F/2 Agar Slant Preparation

With silicate (500 mL agar will make about 28 slants of 18 mL each using 25 mm tubes):

(1) Weigh out 7.5 g agar.
(2) Add agar to a 1 L flask containing 400 mL distilled water (measure amount as calibrations on flasks are not accurate).
(3) Add stir bar and heat on 5, stirring until agar has dissolved. Watch this as it gets hotter, as it will boil over if just left alone.
(4) Add an additional measured 100 mL distilled water to give total volume of 500 mL.
(5) Cover flask top with foil and autoclave tape strip, and autoclave on slow exhaust.
(6) Allow autoclave to cool to 80 °F before opening.
(7) Have sterile 25 mm tubes in a rack in hood with tape moved to side and caps loose.
(8) Also have an extra empty rack for 25 mm tubes and a stirrer/hot plate. If you work quickly the hot plate will not be needed.
(9) Label extra rack with media type and date, then position rack so tubes will slant at the desired angle (two sharpie pens placed under front edge of rack works well).
(10) When agar has cooled but is still liquid and can be held in your hand, aseptically add 0.5 mL Essential Nutrients 1, 0.25 mL Essential Nutrients 2, 0.33 mL concentrated silicate, and 0.5 mL F/2 concentrate + vitamins.
(11) If F/2 without silicate is desired just leave out the silicate above.
(12) If using stirrer/hot plate keep on low to prevent solidification. Keep temperature as low as possible as excess heat destroys vitamins.

(13) With a sterile 25 mL pipette aseptically add 18 mL agar liquid to each sterile tube.

(14) Cap loosely and put in slanted rack.

(15) Allow to solidify for several hours or overnight.

(16) Leave slants out at room temperature for 2–3 days then check for sterility and allow to dry somewhat. Once they are deemed sterile, tubes can be capped tightly and put in rack in refrigerator.

E.5   ALGAE FLASK CULTURE

E.5.1   Back-up Schedule

Healthy stock algae cultures in flasks should be transferred to a new flask with fresh media every 2 weeks. Flask-to-flask transfer should be limited to f generations so new flask cultures should be started directly from slants every 2 months. If flask cultures do not look good it may be necessary to start new cultures from slants earlier. Flask cultures are used for starting fresh carboys so make sure to have enough flask cultures for each species in culture to accommodate both the carboy back-up schedule and the flask back-up schedule.

E.5.2   Flask Culture Maintenance

Flask cultures should be stirred vigorously twice per day to re-suspend any settled algae.

E.5.3   Back-up Protocol (see video)

Stepping Up from Slant Culture to Flask Culture
Starting 500 mL and 1000 mL Flask Cultures from a Slant
All of the following should be done aseptically under the laminar flow hood:

(1) Start with a bottle of sterile water (1 L or 500 mL)

(2) Add to the 1 L bottle of sterile water, 1 mL of concentrated F/2 + vitamins (500 mL bottle will get 0.5 mL of same concentrate stock) stock solution.

(3) Add 1 mL stock sterile Essential Nutrients 1 (EN1) and 0.5 mL sterile Essential Nutrients 2 (EN2).

(4) If you need silicate add 1 mL of sterile stock silicate solution.

(5) Aseptically remove the foil covering the stopper, aseptically place a sterile 1 mL pipette through the hole in the stopper and insert into the flask until the tip touches the bottom of the flask. Take care not to touch the top of the stopper with anything until this is done.

(6) Aseptically remove the stopper with bubbling tube and pour in at least 300 mL (for 500 mL flasks) or 700 mL (for 1000 mL flasks) of whatever nutrient solution you want to use. Replace the stopper and bubbling tube. Do this for each culture you want to inoculate. Label each flask.

(7) To transfer from a stock slant culture use a sterile loop and get as much of the culture in the large loop as possible then place the loop into the media in the flask and stir around to remove as much of the culture as possible. You can also try rubbing the loop end against the side of the flask below the media level to try and remove more culture from the loop.

(8) Put the flask in an incubator and aerate.

Starting 500 mL and 1000 mL Flask Cultures from a Flask

To transfer from a stock flask culture to a new flask, swirl the stock culture to make sure all the culture is in solution and none remains on the bottom of the flask. With a sterile pipette remove 5 mL of stock culture and add to media in the new flask. You should be able to see a faint green tint to the media in the new flask (if you can't see green color add more of the flask culture, 5 mL at a time until you see a faint green color in the new flask). It is always better to add too much rather than too little.

(1) Swirl the new flask to mix well and then put the flask in the small incubator and attach the bubbling tube to one of the empty air tubes in the incubator. Unclip the closure on the tube to allow air to flow and bubbling to begin.

(2) Once all new flasks are inoculated, placed in the incubator and bubbling, place a piece of cardboard on the top shelf to shield the new cultures from the intense light. Leave the cardboard in place for 3–4 days until the culture has had a chance to begin growing and has started to darken slightly, then remove the cardboard and allow cultures to continue to grow until desired density is achieved.

E.6   CARBOY CULTURE

E.6.1   *Back-up Schedule*

Healthy stock algae cultures in carboys should be transferred to a new carboy with fresh media every 2 weeks. Carboy-to-carboy transfer should be limited to 4 generations so new carboy cultures should be started directly from flasks every 2 months. If carboy cultures do not look good it may be necessary to start new cultures from flasks earlier.

E.6.2   *Carboy Cleaning, Sterilization, and Inoculation Procedures*

(1) Used carboys should be washed with a brush and soap and water to remove ALL traces of algae. We use a color code system on the carboys. Put only the color indicated algal species in the coded carboys when inoculating to help cut down on cross contamination.

(2) After washing, rinse each carboy 5–6 times with tap water.

(3) Add 25 mL Clorox to the carboy and fill up with water. Water should fill the carboy neck so that when the cap is put on, water will completely contact the cap surface. Put clean cap on carboy and let sit overnight. Carboys can be filled this way and sit for some time if not needed immediately.

(4) To use carboy for algal culture place in hood and using aseptic technique remove 4400 mL water/Clorox solution from the carboy to give you a total volume of 15 L. Replace cap to maintain a sterile carboy. This can just be rested on top and not snapped down.

(5) Weigh out 3.86 g sodium thiosulfate and aseptically add to the carboy.

(6) Using aseptic technique add one tube of sterile F/2 + nutrients required for the species to be grown to the carboy.

(7) Aseptically place a sterile bubbling tube in the carboy. Large sterile forceps will be needed to push the stopper down into the neck of the carboy. Add a sterile inline filter to the top of the bubbling tube, being sure to get the inlet side up.

(8) Remove carboy from hood, add a small piece of Tygon tubing appropriately sized to fit tube from air and CO_2 source, connect bubbling tube to an air source and bubble air through carboy for at least 1 hour.

(9) Put carboy back in hood and using aseptic technique add inoculum and replace bubbling tube.

(10) Place carboy on light rack and connect bubbling tube to CO_2 source and allow to grow until desired density is reached.

E.6.3   Concentrated Media Tubes for Use in Carboys

(1) Autoclave full rack of 25 mm test tubes with caps (caps should be loose and a piece of autoclave tape over cap and tube with date on tape).

(2) When cool, place in clean hood and using aseptic technique add 15 mL sterile F/2 concentrate + vitamins to each tube.

(3) Add 15 mL of Essential Nutrients 1 (EN1).

(4) Add 7.5 mL of Essential Nutrients 2 (EN2).

(5) If F/2 + vitamins + essential nutrients + silicate is desired, using aseptic technique add an additional 10 mL sterile silicate concentrate to each tube. Re-cap tightly and invert several times to mix well.

(6) Label rack with contents of tubes and date prepared. Make sure label indicates all ingredients in the tubes. You can use one of the large blue test tube racks and have several types of tubes stored in one rack as long as the tubes are separated by a blank row and the rack is well labeled.

(7) Store rack in refrigerator in the dark to help maintain vitamins.

There will be precipitate in the tubes with silicate so be sure to mix well before using.

E.6.4   Carboy Bubbling Tube Preparation

(1) Bubbling tubes can be made by cutting a piece of glass tubing 23 inches long. The tubing we currently have is 48 inches and both ends are smooth. Measure from each end of the long tube and mark 23 inches with a sharpie. Cut out the middle to make 2 tubes approximately 23 inches long and use the smooth tip for the up end of the bubbling tube and the rough cut end for the down end.

(2) Add a 3 inch Styrofoam stopper in which a hole has been made to accommodate the glass tube.

(3) Add a short piece of thick-walled Tygon tubing to the top of the tube. This is used to attach the sterile inline filter for the air source.

(4) Make up 6 tubes.

(5) Using the self-sealing large autoclave bags place 3 tubes in the first bag, stopper end down and together to one side of bag. Then place the other

3 tubes stopper end down in the second bag, again keep ends together in same area of the bag.

(6) Turn one bag over so the blue plastic side lines up with the paper side of the other bag. Push the down ends of tubes into the opposite bag making sure the three down tubes from one bag are in opposite corners of the bags.

(7) Slide bags together so self-sealing strips overlap the opposite bag.

(8) Carefully (slowly/gradually) remove self-sealing strip and seal it against the opposite bag.

(9) Turn the bags over and do the same to the other side.

(10) Add a small strip of autoclave tape to the edges of the seals.

(11) Date bag, and autoclave.

(12) Store in cabinet until needed, being sure to place new bags in the back and oldest bags in the front of the cabinet so oldest stock is used first.

E.6.5 Adding Stopper/Tube Assembly to Each Carboy

(1) Wipe around carboy cap and upper neck with alcohol. Remove cap on carboy and rest on top of carboy.

(2) Carefully open one end of the sterile tube pack. Remove one stopper/tube assembly from pack (touch only on the top portion of the tube assembly to keep sterile and avoid touching within 1–2 inches of the stopper so this area remains sterile should the depth of the tube need to be adjusted once it is placed in the carboy) and without touching the top of the carboy lower the tube into the carboy until the stopper sits on the carboy top.

(3) Using the sterile long forceps push the stopper into the carboy. From the top using a gentle twisting motion, being careful not to touch the sides of the stopper or the insides around the carboy top, push the stopper down into the carboy neck. Just do the best you can as this is difficult to do without touching something.

(4) Now adjust the glass tube in the stopper so it is just off the bottom of the carboy somewhere around the side edge with the stopper well down in the carboy neck.

(5) Attach a sterile in-line filter to the piece of Tygon tubing on the end of the bubbling tube. Be sure the inlet side is up and do not touch the inner edges of the Tygon tubing when inserting the filter.

(6) Make sure the Tygon tubing is snugly attached to the glass bubbling tube. Autoclaving tends to expand the tubing.

(7) Place a short piece of Tygon tubing on the inlet side of the in-line filter to attach the aeration line.

(8) Repeat the procedure until the bubbling tubes at one end of the pack are used.

(9) Be very careful not to touch the glass tubes from the other bubbling tubes.

(10) Fold over the end of the pack and tape it closed, making sure the glass ends of the remaining tubes are protected.

(11) Invert the pack and open the other end; remove the tubes as needed and place in carboys as described above.

E.6.6 Addition of Inoculum

Inoculum should be added aseptically by pouring 1000 mL of heavy suspended cells into a sterile 1000 mL beaker and then pouring the inoculum into a carboy. This should be done in the hood to maintain sterility. Avoid dribbling the inoculum down the sides of the carboy or beaker as much as possible when pouring from the beaker. The hood is limited with respect to space, which is why it is preferable to put the stopper assemblies all in the carboys first and then inoculate. Otherwise you have packets of autoclaved tubes, uninoculated carboys, and a carboy containing inoculum in the hood at the same time. A full carboy of heavy inoculum will inoculate 16 carboys.

E.6.7 Carboy Culture Maintenance

Carboy cultures should be stirred vigorously twice per day to re-suspend any settled algae. CO_2 flow rates should be checked periodically to make sure the CO_2 tank is not empty.

E.6.8 Adjusting Air/CO_2 Flow and Lighting

Place inoculated carboys somewhere where they will receive adequate light and aeration. Adjust the rate of aeration so that bubbling is moderate to vigorous. Light should be reduced until the cells have a chance to acclimate to the new growing conditions. Light can be increased after a day or so. Sometimes cultures will bleach out if they are given too much light too soon and this is often a function of inoculum density.

If CO_2 is needed for faster growth, it can be added at this time. Make sure all gas bottles are chained securely to the wall or other secure structure.

Glossary

A

acquired immunity: a process by which a host rejects a larval mussel infestation before metamorphosis is complete as a result of repeated exposure to mussel larvae.

adductor muscle: large muscles attached to both valves of the shell that close the shell tightly when contracted.

anterior: the shorter end of the shell as measured from the umbo; also considered the front end.

augmentation: releasing a species into an area where that species already occurs in hopes of increasing the long-term viability of the extant population.

B

beak: the raised part of the dorsal margin of the shell; also called the umbo.

beak cavity: the depression or pocket on the inside of each valve leading into the beak.

beak sculpture: the raised loops, ridges, or bumps on the umbo.

Bernoulli effect: this principle states that an increase in the speed of a fluid occurs simultaneously with a decrease in pressure.

bivalve: a mollusk with a shell consisting of two symmetrical valves.

bradytictic: long-term brooders or gravid female mussels that typically brood larvae in the gills over the winter.

byssal thread: one or more fibers secreted by juvenile mussels that help prevent downstream movement by anchoring the mussel to a substrate such as large sand grains, rocks, etc.

C

calcium carbonate: the primary mineral that comprises the shell of a mollusk.

chevron: a V-shaped marking on a mussel shell.

compressed: flattened or pressed together laterally.

conglutinate: a mass of glochidia held together by mucus that often-times resembles a prey item of the host.

contaminant: an impurity in the environment that may be toxic to sensitive organisms.

ctenidia: the respiratory organ found in many mollusks (see gill).

D

demibranch: one half of the paired gills found in bivalves. Each pair is comprised of an inner demibranch and outer demibranch.

direct development: metamorphosis from the larval stage to the juvenile stage inside the marsupium of the female mussel.

distal: away from the center or origin.

dorsal: the top part of the shell where the hinge is located.

E

ectobranchus: female mussels that brood the larvae in the outer demibranch only.

elliptical: having the form of an ellipse, or oval.

elongated: long or extended.

endangered: a species that is threatened with extinction throughout all or a significant portion of its range.

encapsulation: the process by which the host tissue surrounds the larval parasite.

endobranchus: female mussels that brood the larvae in the inner demibranch only.

excurrent aperture: the portion of the mantle where water and waste products are expelled from a freshwater mussel.

extinct: a species that no longer exists.

extirpated: eliminated from a particular area, but still existing in another location.

F

foot: muscular organ used for locomotion in bivalve mollusks.

G

genetic swamping: a process by which a set of naturally evolved, regionally specific genes in a population are lost through hybridization with domesticated varieties.

gill: the respiratory organ in aquatic mollusks also used to brood larvae in some freshwater mussels.

glochidium: the larva stage of some species of freshwater mussel that generally completes metamorphosis to the juvenile stage by parasitizing a host.

gravid: pregnant, brooding eggs or larvae.

growth lines: darkened lines on the surface of the shell that indicate periods of reduced growth of the shell.

H

harvest: the process of extracting larvae from a gravid female.

haustoria: a structure that grows into or around another structure to absorb water or nutrients.

hemacytometer: A device used in the medical industry for counting blood cells that also can be used to estimate algal concentration.

hermaphrodite: an animal or plant having both male and female reproductive organs.

hinge: the elastic part of the shell that unites the valves along the dorsal margin of the shell.

I

impoundment: a free-flowing river that has been converted to a reservoir with a man-made structure.

in vivo: taking place in a living organism.

in vitro: taking place outside a living organism (i.e. test tube or culture plate).

inbreeding depression: reduced fitness in a population as a result of breeding between closely related individuals.

incurrent aperture: the portion of the mantle where water and food enter a freshwater mussel.

infestation: the state of being invaded by a pest or parasite. Freshwater mussel glochidia are parasites so this term is often used interchangeably with inoculation when referring to the process of artificially attaching glochidia to the host species (see inoculation).

inflated: swollen or expanded.

innate immunity: a process by which a host rejects a larval mussel infestation before metamorphosis is complete as a result of an in-born immune response.

inoculation: the introduction of a pathogen or antigen into a living organism to stimulate the production of antibodies. Mussel glochidia are parasites that do stimulate the production of antibodies, so this term is often used interchangeably with infestation when referring to the process of artificially attaching glochidia to the host species (see infestation).

interdentum: a flattened area between the pseudocardinal teeth and lateral teeth in bivalve mollusks.

J

juvenile: the life stage after the larval stage but before the adult stage.

L

larva: the newly hatched, immature form of an animal that undergoes metamorphosis, differing markedly in form or appearance from the adult.

lasidium: the parasitic larval stage in the family Mycetopodidae.

lateral teeth: elongated teeth along the hinge line of the shell of a bivalve mollusk.

left valve: the left half of the shell when the dorsal edge or hinge is facing up and the anterior end is directed forward (away from the collector).

M

mantle: the tissue lining the inside of a mussel shell that encloses the viscera and secretes new shell material from its edges for continued shell growth.

mantle lure: a modified portion of the mantle tissue used to attract a suitable fish host.

mantle magazine: a modified portion of the mantle tissue that serves to both hold small amounts of mature larvae and also attract a suitable fish host.

marsupium: the portion of the gills of a female mussel that brood glochidia.

metamorphosis: a significant change in the form or structure of an animal occurring after birth or hatching. This term is sometimes interchanged with transformation in the freshwater mussel community.

micrometer: one-thousandth of a millimeter.

mollusk: a member of the phylum Mollusca.

muscle scar: the area of attachment of a muscle to the inside of the shell.

mussel bed: a dense, natural aggregation of mussels which can support a diverse variety of benthic fauna.

N

nacre: the inner-most layer of shell that often has a pearl-like appearance.

O

oblong: having the shape of or resembling a rectangle or ellipse.

Ortmann's Law of Stream Position: a species of mussels can have different appearance depending on where they live in a river system (i.e. upstream vs. downstream).

outbreeding depression: reduced fitness in a population as a result of breeding between genetically distant individuals.

ovate: egg-shaped.

ovisac: membranous capsule containing glochidia.

P

pallial line: a linear depression on the inside of the shell.

parasite: an organism that obtains food and shelter from another organism, but contributes nothing to the survival of that organism.

periostracum: the outside layer or covering of the shell.

posterior ridge: a ridge on the back half of the valve running from the umbo to the posterior ventral edge.

posterior slope: the area along the dorsal part of the shell between the posterior ridges of the valves.

propagation: human-supervised plant or animal breeding.

pseudocardinal teeth: the triangular, often serrated, teeth located on the anterior-dorsal part of the shell.

pustule: a bump or raised knob on the outside surface of the shell.

Q

quadrate: square.

R

recruitment: the addition of new members to a wild population through reproduction or immigration.

re-introduction: releasing a species into an area where that species occurred historically but currently does not occur in hopes of establishing new populations.

right valve: the right half of the shell when the dorsal edge or hinge is facing up and the anterior end is directed forward (away from the collector).

S

sexual dimorphism: a condition in which males and females of the same species are morphologically different. In freshwater mussels this is usually indicated by an expanded posterior marsupial area in the female in contrast to a more pointed or bluntly rounded area in the male.

structural eggs: eggs that do not develop but provide color and structure to a conglutinate.

subadult: an individual that has passed through the juvenile period but not yet attained typical adult characteristics.

sulcus: a shallow depression or furrow on the outside surface of the shell.

T

tachytictic: short-term brooders or gravid females that typically brood larvae in the gills only during the summer.

tetragenous: female mussels that brood the larvae in both the inner and the outer demibranchs.

transform: a complete or major change in appearance, form, etc. This term is sometimes used interchangeably with metamorphosis in the freshwater mussel community.

truncated: having the end shortened or squared off.

tubercle: a small lump or nodule on the outer surface of a shell.

U

umbo: the oldest part of a bivalve shell (also called a beak) located in the anterio-dorsal portion of the shell in most freshwater mussels.

unionid: freshwater mussels in the order Unionoida. The prefix "unio" means pearl in latin.

V

valve: one of the two halves of the shell of a bivalve mollusk.

ventral: the bottom edge of the shell.

W

water tube: internal segments of the gill created by lamellae and extending from the dorsal margin to the ventral margin that are used to brood larvae and mold conglutinates.

References

Aji, L.P. 2011. The use of algae concentrates, dried algae and algal substitutes to feed bivalves. *Makara of Science Series* 15:1–8.

Akiyama, Y. and T. Iwakuma. 2007. Survival of glochidial larvae of the freshwater pearl mussel, *Margaritifera laevis* (Bivalvia: Unionoida), at different temperatures: a comparison between two populations with and without recruitment. *Zoological Science* 24(9):890–893.

Aldridge, D.C. and A.L. McIvor. 2003. Gill evacuation and release of glochidia by *Unio pictorum* and *Unio tumidus* under thermal and hypoxic stress. *Journal of Molluscan Studies* 69:55–59.

Allen, D.C. and C.C. Vaughn. 2011. Density-dependent biodiversity effects on physical habitat modification by freshwater bivalves. *Ecology* 92:1013–1019.

Allen, D.C., M.C. Hove, B.E. Sietman, *et al.* 2007. Early life-history and conservation status of *Venustaconcha ellipsiformis* (Bivalvia, Unionidae) in Minnesota. *American Midland Naturalist* 157:74–91.

Allen, D.C., C.C. Vaughn, J.F. Kelly, J.T. Cooper, and M.H. Engel. 2012. Bottom-up biodiversity effects increase resource subsidy flux between ecosystems. *Ecology* 93:2165–2174.

Allen, D.C., H.S. Galbraith, C.C. Vaughn, and D.E. Spooner. 2013. A tale of two rivers: implications of water management practices for mussel biodiversity outcomes during droughts. *Ambio* 42:881–891.

Anfinson, J.O. 2003. *The River We Have Wrought: A History of the Upper Mississippi*. University of Minnesota Press, Minneapolis, MN, USA.

Aprahamian, M.W., K. Martin Smith, P. McGinnity, S. McKelvey, and J. Taylor. 2003. Restocking of salmonids: opportunities and limitations. *Fisheries Research* 62:211–227.

Araujo, R. and M.A. Ramos. 2001. Life history data on the virtually unknown *Margaritifera auricularia*. Pages 143–152 in G. Bauer and K. Wachtler (editors), *Ecology and Evolution of the Freshwater Mussels Unionoida*. Ecological Studies 145. Springer-Verlag, Berlin, Germany.

Araujo, R., N. Camara, and M.A. Ramos. 2002. Glochidium metamorphosis in the endangered freshwater mussel *Margaritifera auricularia* (Spengler, 1973): a histological and scanning electron miscroscopy study. *Journal of Morphology* 254:259–265.

Arey, L.B. 1921. An experimental study on glochidia and the factors underlying encystment. *Journal of Experimental Zoology* 38:463–499.

1932. A microscopical study of glochidia immunity. *Journal of Morphology* 53:367–379.

Atkinson, A.J., P.C. Trenham, R.N. Fisher, *et al.* 2004. *Designing Monitoring Programs in an Adaptive Management Context for Regional Multiple Species Conservation Plans.* 74pp.

Atkinson, C.L. and C.C. Vaughn. 2015. Biogeochemical hotspots: temporal and spatial scaling of the impact of freshwater mussels on ecosystem function. *Freshwater Biology* 60:563–574.

Atkinson, C.L., C.C. Vaughn, K.J. Forshay, and J.T. Cooper. 2013. Aggregated filter-feeding consumers alter nutrient limitation: consequences for ecosystem and community dynamics. *Ecology* 94:1359–1369.

Atkinson, C.L., J.P. Julian, and C.C. Vaughn. 2014a. Species and function lost: role of drought in structuring stream communities. *Biological Conservation* 176:30–38.

Atkinson, C.L., J.F. Kelly, and C.C. Vaughn. 2014b. Tracing consumer derived nitrogen in riverine food webs. *Ecosystems* 17:485–496.

Augspurger, T., A.E. Keller, M.C. Black, G. Cope, and F.J. Dwyer. 2003. Derivation of water quality guidance for protection of freshwater mussels (Unionidae) from ammonia exposure. *Environmental Toxicology and Chemistry* 22:2569–2575.

Baggett, L.P., S.P. Powers, R. Brumbaugh, *et al.* 2014. *Oyster Habitat Restoration Monitoring and Assessment Handbook.* The Nature Conservancy, Arlington, VA, USA. 96pp.

Barfield, M. and G.T. Watters. 1998. Non-parasitic life cycle in the green floater, *Lasmigona subviridis* (Conrad, 1835). *Triannual Unionid Report* 16:22.

Barnhart, M.C. 1997. *Fertilization Success in Freshwater Mussels.* Report to the Missouri Department of Conservation, Columbia, MO, USA.

2001. *Fish Hosts and Culture of Mussel Species of Special Concern.* Report to the Missouri Department of Conservation, Columbia, MO, USA.

2003. *Culture and Restoration of Mussel Species of Concern.* Final Report. Missouri Department of Conservation and US Fish and Wildlife Service, Columbia, MO, USA.

2004. Propagation and restoration of mussel species of concern. *Endangered Species Grant Interim Report* (E-1–42).

2006. Buckets of muckets: a compact system for rearing juvenile freshwater mussels. *Aquaculture* 254:227–233.

Barnhart, M.C. and A.D. Roberts. 1997. Reproduction and fish hosts of unionids from the Ozark Uplifts. Pages 15–20 in K.S. Cummings, A.C. Buchanan, T.J. Naimo, and C.A. Mayer (editors), *Conservation and Management of*

Freshwater Mussels II: Initiatives for the Future. Proceedings of a UMRCC Symposium, 16–18 October 1995, St. Louis, MO. Upper Mississippi River Conservation Committee, Rock Island, IL, USA.

Barnhart, M.C., W.R. Haag, and W.N. Roston. 2008. Adaptations to host infection and larval parasitism in Unionoida. *Journal of the North American Benthological Society* 27(2):370–394.

Barnhart, M.C., B. Glidewell, A. Maynard, *et al.* 2015. Pulsed flow-through systems for the laboratory culture of early life stages of freshwater mussels. Presented at *"Rearing of Unionid Mussels,"* Clervaux, Luxembourg. www .heppi.com/seminar/Uc_Lux_2015_Barnhart1.pdf (accessed September 2017).

Bauer, G. 1986. The status of the freshwater pearl mussel *Margaritifera margaritifera* L. in the south of its European Range. *Biological Conservation* 38:1–9.

1987. Reproductive strategy of the freshwater pearl mussel *Margaritifera margaritifera. Journal of Animal Ecology* 56:691–704.

1988. Threats to the freshwater pearl mussel, *Margaritifera margaritifera* L. in central Europe. *Biological Conservation* 45:239–253.

1992. Variation in the life span and size of the freshwater pearl mussel. *Journal of Animal Ecology* 61:425–436.

Bayne, B.L., D.L. Holland, M.N. Moore, D.M. Lowe, and J. Widdows. 1978. Further studies on the effects of stress in the adult on the eggs of *Mytilus edulis. Journal of the Marine Biological Association, UK* 58:824–841.

Beck K.M. and R.J. Neves. 2003. An evaluation of selective feeding by three age-groups of the rainbow mussel *Villosa iris. North American Journal of Aquaculture.* 65:203–209.

Benke, A. 1990. A perspective on America's vanishing streams. *Journal of the North American Benthological Society* 9(1):77–88.

Berenberg, C.J. and G.W. Patterson. 1981. The relationship between dietary phytosterols and the sterols of wild and cultivated oysters. *Lipids* 16:276–278.

Berka, R. 1986. *The Transport of Live Fish: A Review.* EIFAC Technical Paper (48):52pp.

Blystad, C.N. 1923. Significance of larval mantle of fresh-water mussels during parasitism, with notes on a new mantle condition exhibited by *Lampsilis luteola. Bulletin of the United States Bureau of Fisheries* 39:203–219.

Bogan, A.E. 2008. Global diversity of freshwater mussels (Mollusca, Bivalvia) in freshwater. *Hydrobiologia* 595:139–147.

Bouchet, P., J.-P. Rocroi, R. Bieler, J.G. Carter, and E.V. Coan. 2010. Nomenclator of bivalve families: classification of bivalve families. *Malacologia* 52(2): 1–184.

Boyer, S.L., A.A. Howe, N.W. Juergens, and M.C. Hove. 2011. A DNA-barcoding approach to identifying juvenile freshwater mussels (Bivalvia:Unionidae) recovered from naturally infested fishes. *Journal of the North American Benthological Society* 30:182–194.

Brady, T.R., D. Aloisi, R. Gordon, and G. Wege. 2012. A method for culturing mussels using in-river cages. *Journal of Fish and Wildlife Management* 2(1):85–89.

Brown, M.R., S.W. Jeffrey, J.K. Volkman, and G.A. Dunstan. 1997. Nutritional properties of microalgae for mariculture. *Aquaculture* 151:315–331.

Burkhead, N.M. 2012. Extinction rates in North American freshwater fishes, 1900–2010. *Bioscience* 62(9):798–808.

Busack, C.A. and K. P. Currens. 1995. Genetic risks and hazards in hatchery operations: fundamental concepts and issues. Pages 71–80 in H. L. Schramm, Jr. and R. G. Piper (editors), *American Fisheries Society Symposium 15.* American Fisheries Society, Bethesda, MD, USA.

Bush, A.L. 2008. *An Assessment of Suitable Feed Quantity and Quality for Riffleshell Mussels (Epioblasma spp.) Held in Captivity.* Master's thesis. Virginia Polytechnic Institute and State University, Blacksburg, VA, USA.

Byrne, R.A. and B.R. McMahon. 1991. Acid–base and ionic regulation, during and following emersion, in the freshwater bivalve, *Anodonta grandis simpsoniana* (Bivalvia: Unionidae). *Biological Bulletin of the Marine Biology Laboratory, Woods Hole* 181:289–297.

Canessa, S., D. Hunter, M. McFadden, G. Marantelli and M.A. McCarthy. 2014. Optimal release strategies for cost-effective reintroductions. *Journal of Applied Ecology* 51:1107–1115.

Carrizo, S. F., K. G. Smith, and W. R. T. Darwall. 2013. Progress towards a global assessment of the status of freshwater fishes (Pisces) for the IUCN Red List: application to conservation programmes in zoos and aquariums. *International Zoo Yearbook* 47(1):46–64.

Carus, C.G. 1832. Neue Untersuchungen uber die Entwicklungsgeschichte unserer Flubmuschel. *Verh K Leopoldinisch-Carolinischen Akad Naturforsch* 16:1–81.

Chen, L.Y., A.G. Heath, and R.J. Neves. 2001. An evaluation of air and water transport of freshwater mussels (Bivalvia: Unionidae). *American Malacological Bulletin* 16:147–154.

Clavijo, C., A. Carranza, F. Scarabino, and A. Southullo. 2010. Conservation priorities for Uruguayan land and freshwater molluscs. *Tentacle* 18:14–16.

Coker, R.E. and T. Surber. 1911. A note on the metamorphosis of the mussel *Lampsilis laevissimus. Biological Bulletin* 20:179–182.

Coker, R.E., A.F. Shira, H.W. Clark, and A.D. Howard. 1921. Natural history and propagation of fresh-water mussels. *Bulletin of the Bureau of Fisheries* 37:75–181.

Conner, C.H. 1909. Supplementary notes on the breeding seasons of the Unionidae. *Nautilus* 22:111–112.

Converse, S. J., C. T. Moore, and D.P. Armstrong. 2013. Demographics of reintro-duced populations: Estimation, modeling and decision analysis. *The Journal of Wildlife Management* 77(6):1081–1093.

Cope, W.G., T.J. Newton, and C.M. Gatenby. 2003. Review of techniques to prevent introduction of zebra mussels (*Driessena polymorpha*) during native mussel (Unionidea) conservation activities. *Journal of Shellfish Research* 22(1):177–184.

Cox, I.G. (1994) Stocking strategies. *Fisheries Management and Ecology* 1:15–30.

Cummings, K.S. and D.L. Graf. 2010. Mollusca: Bivalvia. Pages 309–384 in J.H. Thorp and A.P. Covich (editors), *Ecology and Classification of North American Freshwater Invertebrates. 3rd edn.* Academic Press, San Diego, CA, USA.

Cummings, K.S. and C.A. Mayer. 1993. *Distribution and Host Species of the Federally Endangered Freshwater Mussel, Potamilus capax (Green, 1832) in the Lower Wabash River, Illinois and Indiana.* Technical Report 1993:1–29.

Cummings, K.S., L. Suloway, and L.M. Page. 1988. *The Freshwater Mussels (Mollusca: Bivalvia: Unionidae) of the Embarras River in Illinois: Thirty Years of Stream Change.* Illinois Natural History Survey Technical Report 1988.

Davis, G.M. and S.L.H. Fuller. 1981. Genetic relationships among Recent Unionacea (Bivalvia) of North America. *Malacologia* 20:217–253.

Dickinson, B.D. and B.E. Sietman. 2008. Recent observations of metamorphosis without parasitism in *Utterbackia imbecillis. Ellipsaria* 10(1):7–8.

Dietz, T.H. 1979. Uptake of sodium and chloride by freshwater mussels. *Canadian Journal of Zoology* 57:156–160.

Distler, D.A. and D.E. Bleam. 1995. Decline in the diversity of Bivalvia, Ninnescah River, Kansas. *Transactions of the Kansas Academy of Science* 98:156–159.

Downing, J. A., P. Van Meter, and D. A. Woolnough. 2010. Suspects in evidence: a review of the causes of extirpation and decline in freshwater mussels. *Animal Biodiversity and Conservation* 33(2):151–185.

Dudgeon, D. 1999. *Tropical Asian Streams: Zoobenthos, Ecology and Conservation.* Hong Kong University Press, Hong Kong, People's Republic of China.

Edgar, A.L. 1965. Observations of the sperm of the pelecypod *Anodontoides ferussacianus* (Lea). *Transactions of the American Microscopical Society* 84:228–230.

Engel, H. 1990. *Untersuchungen zur Autökologie von Unio crassus (PHILIPSSON) in Norddeutschland.* Dissertation, Universität Hanover, Hanover, Germany.

Enright, C. T., G. F. Newkirk, J. S. Craigie, and J. D. Castell. 1986. Evaluation of phytoplankton diets for juvenile *Ostrea edulis. Journal of Experimental Marine Biology and Ecology* 96:1–13.

Etnier, D.A. and W.C. Starnes. 1993. *The Fishes of Tennessee*. University of Tennessee Press, Knoxville, TN, USA.

Evans, R.R. 2001. *Historical and Contemporary Distributions of Aquatic Mollusks in the Upper Canasauga River System of Georgia and Tennessee*. Thesis, University of Tennessee, Chattanooga, TN, USA.

Eybe, T., F. Thielen, T. Bohn, and B. Sures. 2013. The first millimetre: rearing juvenile freshwater pearl mussels (*Margaritifera margaritifera* L.) in plastic boxes. *Aquatic Conservation: Marine and Freshwater Ecosystems*. 23:964–975.

Ferguson, C.F., M.J. Blum, M. Raymer, M.S. Eakles, and D.E. Krane. 2013. Population structure, multiple paternity and long-distance transport of spermatozoa in the Near Threatened freshwater mussel *Lampsilis cardium* (Bivalvia: Unionidae). *Freshwater Science* 32(1):267–282.

Fisher, G.R. and R.V. Dimock, Jr. 2000. Viability of glochidia of *Utterbackia imbecillis* following their removal from the parental mussel. Pages 185–188 in *Freshwater Mollusk Symposium Proceedings*. Ohio Biological Survey, Columbus, OH, USA.

Fleischauer-Rossing, S. 1990. *Untersuchungen zur Autokologie von Unio tumidus Philipsson und Unio pictorum L. (Bivalvia) unter besonderer Berucksichtigung der fruhen postparasitaren Phase*. Thesis, Universität Hanover, Hanover, Germany.

Francis-Floyd, R. 1995. *The Use of Salt in Aquaculture*. Cooperative Extension Service Fact Sheet VM 86. University of Florida, Gainesville, FL, USA.

Fritts, A.K., M.C. Barnhart, M. Bradley, *et al*. 2014. Assessment of toxicity test endpoints for freshwater mussel larvae (glochidia). *Environmental Toxicology and Chemistry* 33(1):199–207.

Fritts, M.W., A.K. Fritts, S.A. Carleton, and R.B. Bringolf. 2013. Shifts in stable-isotope signatures confirm parasitic relationship of freshwater mussel glochidia attached to host fish. *Journal of Molluscan Studies* 79(2):163–167.

Fryer, G. 1961. The developmental history of *Mutela bourguignati* (Ancey) Bourguignat (Mollusca: Bivalvia). *Philisophical Transactions of the Royal Society of Edinburgh* 244:259–298.

Fuller, S.L.H. 1974. Clams and mussels (Mollusca: Bivalvia). Pages 215–273 in C. W. Hart and S. L. H. Fuller (editors), *Pollution Ecology of Freshwater Invertebrates*. Academic Press, Inc., New York, USA.

Fustish, C.A. and R.E. Millemann. 1978. Glochidiosis of salmonid fishes. II. Comparison of tissue response of Coho and Chinook salmon to experimental infection with *Margaritifera margaritifera* (L.) (Pelecypoda: Margaritanidae). *Journal of Parasitology* 64(1):155–157.

Gatenby, C.M. 1994. *Development of a Diet for Rearing Juvenile Freshwater Mussels*. Master's Thesis, Virginia Polytechnic Institute and State University, Blacksburg, VA, USA.

————. 2000. *A Study of Holding Conditions, Feed Ration, and Algal Foods for the Captive Care of Freshwater Mussels*. PhD. Dissertation. Virginia Polytechnic Institute and State University, Blacksburg, VA, USA.

Gatenby, C.M., R.J. Neves, and B.C. Parker. 1996. Influence of sediment and algal food on cultured juvenile freshwater mussels. *Journal of the North American Benthological Society* 15:597–609.

Gatenby, C.M., B.C. Parker, and R.J. Neves. 1997. Growth and survival of juvenile rainbow mussels, *Villosa iris* (Lea, 1829) (Bivalvia: Unionidae), reared on algal diets and sediment. *American Malacological Bulletin* 14:57–66.

Gatenby, C.M., P.A. Morrison, R.J. Neves, and B.C. Parker. 2000. A protocol for the salvage and quarantine of unionid mussels from zebra mussel-infested waters. Pages 9–18 in R.A. Tankersley, D.I. Warmolts, G.T. Watters, B. J. Armitage, P.D. Johnson, and R.S. Butler (editors), *Freshwater Mollusk Symposia Proceedings*. Ohio Biological Survey, Columbus, OH, USA.

Gatenby, C.M., D.M. Orcutt, D.A. Kreeger, *et al.* 2003. Biochemical composition of three algal species proposed as food for captive freshwater mussels. *Journal of Applied Phycology* 15:1–11.

Gatenby, C.M., D.A. Kreeger, M.A. Patterson, M. Marinni, and R.J. Neves. 2013. Clearance rates of *Villosa iris* (Bivalvia:Unionidae) fed different rations of the alga *Neochloris oleoabundans*. *Walkerana*, 16(1):9–20.

Geist, J. 2010. Strategies for the conservation of endangered freshwater pearl mussels (*Margaritifera margaritifera* L.): a synthesis of conservation genetics and ecology. *Hydrobiologia* 644:69–88.

George, A.L., B.R. Kuhajda, J.D. Williams, *et al.* 2009. Guidelines for propagation and translocation for freshwater fish conservation. *Fisheries* 31(11):529–545.

Gordon, M.E. and D.G. Smith. 1990. Autumnal reproduction in *Cumberlandia monodonta* (Unionoidea: Margaritiferidae). *Transactions of the American Microscopical Society* 109(4):407–411.

Graf, D.L. 2013. Patterns of freshwater bivalve global diversity and the state of phylogenetic studies on the Unionoida, Sphaeriidae, and Cyrenidae. *American Malacological Bulletin* 31(1):135–153.

Graf, D.L. and K.S. Cummings. 2006. Palaeoheterodont diversity (Mollusca: Trigonioida + Unionoida): what we know and what we wish we knew about freshwater mussel evolution. *Zoological Journal of the Linnean Society* 148:343–394.

2013. The Freshwater Mussels (Unionoida) of the World (and other less consequential bivalves), updated 5 August 2015. MUSSEL project web site, www.mussel-project.net (accessed September 2017).

Graf, D.L. and D.Ó. Foighil. 2000. The evolution of brooding characters among the freshwater mussels (Bivalvia: Unionoidea) of North America. *Journal of Molluscan Studies* 66:157–170.

Grobler, P.J., J.W. Jones, N.A. Johnson, *et al.* 2006. Patterns of genetic differentiation and conservation of the slabside pearlymussel, *Lexingtonia dolabelloides* (Lea, 1840) in the Tennessee River drainage. *Journal of Molluscan Studies* 72:65–75.

Haag, W.R. 2002. *Spatial, Temporal, and Taxonomic Variation in Population Dynamics and Community Structure of Freshwater Mussels.* Dissertation, University of Mississippi, Oxford, MS, USA.

2012. *North American Freshwater Mussels: Natural History, Ecology and Conservation.* Cambridge Univerity Press, New York, USA.

Haag, W.R. and A.L. Rypel. 2011. Growth and longevity in freshwater mussels: evolutionary and conservation implications. *Biological Reviews* 86:225–247.

Haag, W.R. and J.L. Staton. 2003. Variation in fecundity and other reproductive traits in freshwater mussels. *Freshwater Biology* 48:2118–2130.

Haag, W.R. and M.L. Warren. 1997. Host fishes and reproductive biology of 6 freshwater mussel species from the Mobile Basin, USA. *Journal of the North American Benthological Society.* 16(3):576–585.

Haag, W.R. and M.L. Warren. 2003. Host fishes and infection strategies of freshwater mussels in large Mobile Basin streams, USA. *Journal of the North American Benthological Society* 22(1):78–91.

Haag, W.R. and M.L. Warren. 2004. *Species Richness and Total Population Size of Freshwater Mussels in Horse Lick Creek, Kentucky in 2003.* Unpublished report, USDA Forest Service, Oxford, MS, USA.

Haag, W.R. and J.D. Williams. 2014. Biodiversity on the brink: an assessment of conservation strategies for North American freshwater mussels. *Hydrobiologia* 735:45–60.

Hadfield, C.A. and L.A. Clayton. 2011. Fish quarantine: current practices in public zoos and aquaria. *Journal of Zoo and Wildlife Medicine* 42(4):641–650.

Haggerty, T.M., J.T. Garner, and R.L. Rogers. 2005. Reproductive phenology in *Megalonaias nervosa* (Bivalvia: Unionidae) in Wheeler Reservoir, Tennessee River, Alabama, USA. *Hydrobiologia* 539:131–136.

Haggerty, T.M., J.T. Garner, A.E. Crews, and R. Kawamura. 2011. Reproductive seasonality and annual fecundity in *Arcidens confragosus* (Unionidae: Unioninae: Anodontini) from Tennessee River, Alabama, USA. *Invertebrate Reproduction and Development* 55(4):230–235.

Halverson, A. 2010. *An Entirely Synthetic Fish: How Rainbow Trout Beguiled America and Overran the World*. Yale University Press, New Haven, KT, USA.

Hanlon, S.D., M.A. Petty, and R.J. Neves. 2009. Status of native freshwater mussels in Copper Creek, Virginia. *Southeastern Naturalist* 8:1–18.

Hartfield, P.D. and E. Hartfield. 1996. Observations on the conglutinates of *Pychobranchus greenii* (Conrad 1834) (Mollusca: Bivalvia: Unionoidea). *American Midland Naturalist* 135:370–375.

Hastie, L.C. and M. R. Young. 2003a. *Conservation of the Freshwater Pearl Mussel: Captive Breeding Techniques*. Conserving Natura 2000 Rivers Ecology Series No.2, English Nature, Peterborough, UK.

Hastie, L.C. and M.R. Young. 2003b. Timing of spawning and glochidial release in Scottish freshwater pearl mussel (*Margaritifera margaritifera*) populations. *Freshwater Biology* 48:2107–2117.

Heard, W.H. 1998. Brooding patterns in freshwater mussels. *Malacological Review, Supplement 7, Bivalvia*. 1:105–121.

Heard, W.H. and R.H. Guckert. 1970. A re-evaluation of the Recent Unionacea (Pelecypoda) of North America. *Malacologia* 10:333–355.

Heasman, M.P., T.M. Sushames, J.A. Diemar, W.A. O'Connor, and L.A. Foulkes. 2001. *Production of Micro-algal Concentrates for Aquaculture. Part 2: Development and Evaluation of Harvesting, Preservation, Storage and Feeding Technology*. NSW Fisheries Final Report Series No. 34, ISSN 1440-3544.

Heath, A.G. and L.Y. Chen. 1996. *Using Physiological Measurements to Establish Water Quality Criteria for Freshwater Mussels*. Final Report to U.S. Fish and Wildlife Service, Asheville, NC, USA.

Helm, M.M. and N. Bourne. 2004. *Hatchery Culture of Bivalves: A Practical Manual*. FAO, Rome.

Helm, M.M., Holland, D.L. and Stephenson, R.R. 1973. The effect of supplementary algal feeding of a hatchery breeding stock of *Ostrea edulis* L. on larval vigour. *Journal of the Marine Biological Association U.K.* 53:673–684.

Henley, W.F., L.L. Zimmerman, R.J. Neves, and M. Kidd. 2001. Design and evaluation of recirculating water systems for maintenance and propagation of freshwater mussels. *North American Journal of Aquaculture* 63:144–155.

Henley, W.F., P.J. Grobler, and R.J. Neves. 2006. Non-invasive method to obtain DNA from freshwater mussels (Bivalvia: Unionidae). *Journal of Shellfish Research* 25:975–977.

Hochwald, S. 1997. *Das Beziehungsgefüge innerhalb der Größenwachstums- und Fortpflanzungsparameter bayrischer Bachmuschelpopulationen (Unio crassus Phil. 1788) und dessen Abhängigkeit von Umweltparametern*. Dissertation, Universität Bayreuth, Bayreuth, Germany.

Hoeh, W.R., K.S. Frazer, E. Naranjo-García, and R.J. Trdan. 1995. A phylogenetic perspective on the evolution of simultaneous hermaphrodism in a freshwater mussel clade (Bivalvia: Unionidae: Utterbackia). *Malacolgical Review* 28:25–42.

Hoggarth, M. A. 1992. An examination of the glochidia-host relationship reported in the literature for North American species of Unionacea (Mollusca: Bivalvia). *Malacology Data Net* 3:1–30.

Hoggarth, M. A. and A.S. Gaunt. 1988. Mechanics of glochidial attachment (Mollusca: Bivalvia: Unionidae). *Journal of Morphology* 198:71–81.

Hornbach, D.J., C.M. Way, T.E. Wissing, and A.J. Burky. 1984. Effects of particle concentration and season on the filtration rates of the freshwater clam, *Sphaerium striatinum* Lamarck (Bivalvia: Pisidiidae). *Hydrobiologia* 108:83–96.

Hornbach, D.J., V.J. Kurth, and M.C. Hove. 2010. Variation in freshwater mussel shell sculpture and shape along a river gradient. *American Midland Naturalist* 164:22–36.

Hove, M.C., B.E. Sietman, J.E. Bakelaar, *et al.* 2011. Early life history and distribution of Pistolgrip (*Tritogonia verrucosa* (Rafinesque, 1820)) in Minnesota and Wisconsin. *American Midland Naturalist* 165:338–354.

Howard, A.D. 1914. *Experiments in the Propagation of Freshwater Mussels of the Quadrula Group.* Report of the US Commissioner of Fisheries 1913 Appendix 4:1–52. Issued separately as US Bureau of Fisheries Document No. 801. US Government Printing Office, Washington, DC, USA.

1915. Some exceptional cases of breeding among the Unionidae. *Nautilus* 29(1):4–11.

1922. Experiments in the culture of freshwater mussels. *Bulletin of the United States Bureau of Fisheries* 38:63–89.

Howard, A.D. and B.J. Anson. 1922. Phases in the parasitism of the Unionidae. *Journal of Parasitology* 9:68–82.

Howells, R.G., C.M. Mather, and J.A.M. Bergmann. 1997. Conservation status of selected freshwater mussels in Texas. In K.S. Cummings, A.C. Buchanan, C.A. Mayer, and T.J. Naimo (editors), *Conservation and Management of Freshwater Mussels II: Initiatives for the Future. Proceedings of a UMRCC Symposium, Rock Island, Illinois.* UMRCC, Onalaska, WI, USA.

Hruska, J. 1992. The freshwater pearl mussel in South Bohemia: Evaluation of the effect of temperature on reproduction, growth and age structure of the population. *Archiv für Hydrobiologie* 126:181–191.

Imlay, M.J. 1973. Effects of potassium on survival and distribution of freshwater mussels. *Malacologia* 12:97–113.

International Union for Conservation of Nature. 2016. *The IUCN Red List of Threatened Species.* Version 2016-3. www.iucnredlist.org (accessed April 2017).

Ishibashi, R., A. Komaru and T. Kondo. 2000. Sperm sphere in unionid mussels (Bivalvia: Unionidae). *Zoological Science* 17:947–950.

Isom, B.G. 1971. A biologist's look at the history of Muscle Shoals (Mussel Shoals). *Malacological Review* 4:203–206.

Isom, B.G. and R.G. Hudson. 1982. *In vitro* culture of parasitic freshwater mussel glochidia. *Nautilus* 96:147–151.

Jackson, R. B., S. R. Carpenter, C. N. Dahm, *et al.* 2001. Water in a changing world. *Ecological Applications* 11(4):1027–1045.

Jansen, W., G. Bauer, and E. Zahner-Meike 2001. Glochidial mortality in freshwater mussels. Pages 185–211 in G. Bauer and K. Wachtler (editors), *Ecological Studies, Vol. 145, Ecology and Evolution of the Freshwater Mussels Unionoida*, Springer-Verlag, Berlin, Germany.

Jelks, H. L., S. J. Walsh, N. M. Burkhead, *et al.* 2008. Conservation status of imperiled North American freshwater and diadromous fishes. *Fisheries* 33(8):372–407.

Jeong, K.H. 1989. An ultrastructural study on the glochidium and glochidial encystment on the host fish. *Korean Journal of Malacology* 5:1–9.

Johnson, P. D., A. E. Bogan, K. M. Brown, *et al.* 2013. Conservation status of freshwater gastropods of Canada and the United States. *Fisheries* 38(6):247–282.

Jones, H.A., R.D. Simpson, and C.L. Humphrey. 1986. The reproductive cycles and glochidia of freshwater mussels (Bivalvia: Hyriidae) of the Macleay River, northern New South Wales, Australia. *Malacalogia* 27(1):185–202.

Jones, J.W. and R.J. Neves. 2010. Descriptions of a new species and a new subspecies of freshwater mussels, *Epioblasma ahlstedti* and *Epioblasma florentina aureola* (Bivalvia: Unionidae), in the Tennessee River drainage, USA. *Nautilus* 124(2):77–92.

Jones, J.W., R.J. Neves, S.A. Ahlstedt, and R.A. Mair. 2004. Life history and propagation of the endangered dromedary pearlymussel (*Dromus dromas*) (Bivalvia:Unionidae). *Journal of the North American Benthological Society* 23(3):515–525.

Jones, J.W., R.A. Mair, and R.J. Neves. 2005. Factors affecting survival and growth of juvenile freshwater mussels (Bivalvia: Unionidae) cultured in recirculating aquaculture systems. *Journal of North American Aquaculture* 67:210–220.

Jones, J.W., E.M. Hallerman, and R.J. Neves. 2006a. Genetic management guidelines for captive propagation of freshwater mussels (Unionoidea). *Journal of Shellfish Research* 25(2):527–535.

Jones, J.W., R.J. Neves, S.A. Ahlstedt and E.A. Hallerman. 2006b. A holistic approach to taxonomic evaluation of two closely related endangered freshwater mussel species, the oyster mussel *Epioblasma capsaeformis* and tan riffleshell *Epioblasma florentina walkeri* (Bivalvia: Unionidae). *Journal of Molluscan Studies* 72:267–283.

Jones, J.W., R.J. Neves and E.M. Hallerman. 2012. Population performance criteria to evaluate reintroduction and recovery of two endangered mussel species, *Epioblasma brevidens* and *Epioblasma capsaeformis* (Bivalvia: Unionidae). *Walkerana: Journal of the Freshwater Mollusk Conservation Society* 15(1):27–44.

Jones, J.W., S.A. Ahlstedt, B.J.K. Ostby, *et al.* 2014. Clinch River freshwater mussels upstream of Norris Reservoir, Tennessee and Virginia: a quantitative assessment from 2004–2014. *Journal of the American Water Resources Association* 50:820–836.

Jørgensen, C.B. 1990. *Bivalve Filter Feeding: Hydrodynamics, Bioenergetics, Physiology and Ecology*. Olsen and Olsen, Fredensborg, Denmark.

Karna, D.W. and R.E. Millemann. 1978. Glochidiosis of salmonid fishes. III. Comparative susceptibility to natural infection with *Margaritifera margaritifera* (L.) (Pelecypoda: Margaritanidae) and associated histopathology. *Journal of Parasitology* 64:528–537.

Kat, P.W. 1984. Parasitism and the Unionacea (Bivalvia). *Biological Reviews* 59:189–207.

Keller, A.E. and S.G. Zam. 1990. Simplification of *in vitro* culture techniques for freshwater mussels. *Environmental Toxicology and Chemistry*. 9:1291–1296.

Khym, J.R. and J.B. Layzer. 2000. Host suitability for glochidia of *Ligumia recta*. *American Midland Naturalist* 143:178–184.

Knauer, J. and P.C. Southgate. 1999. A review of the nutritional requirements of bivalves and the development of alternative and artificial diets for bivalve aquaculture. *Reviews in Fisheries Science* 7:241–280.

Kondo, T. 1990. Reproductive biology of a small bivalve *Grandiera burtoni* in Lake Tanganyika. *Venus* 49:120–125.

Kreeger, D.A. and C.J. Langdon. 1993. Effect of dietary protein content on growth of juvenile mussels, *Mytilus trossulus* (Gould 1850). *Biological Bulletin* 185:123–139.

Lane, T., H. Dan, and J. W. Jones. 2014. *Reintroduction of Endangered Freshwater Mussel Populations to High Priority Geographic Areas in the Upper Tennessee River System*. Unpublished report, U.S. Fish and Wildlife Service, Asheville, NC, USA.

Lasee, B.A. 1991. *Histological and Ultrastructural Studies of Larval and Juvenile Lampsilis (Bivalvia) from the Upper Mississippi River.* Doctoral Dissertation. Iowa State University, Ames, IA, USA.

Layzer, J.B., M.E. Gordon, and R. M. Anderson. 1993. Mussels: the forgotten fauna of regulated rivers. A case study of the Caney Fork River. *Regulated Rivers: Research and Management* 8:63–71.

Lefevre, G. and W.C. Curtis. 1910. Reproduction and parasitism in the Unionidae. *Journal of Experimental Zoology* 9:79–115.

1912. Studies on the reproduction and artificial propagation of freshwater mussels. *Bulletin of the Bureau of Fisheries* 30:105–201.

Lellis, W.A. and T.L. King. 1998. Release of metamorphosed juveniles by the green floater, *Lasmigona subviridis. Triannual Unionid Report* 16:23.

Leydig, F. 1866. *Mittheilung uder den Parisitismus junger Unioniden an Fischen in Noll.* Inaugural-Dissertation, Tübingen, Frankfurt-am-Main, Germany.

Lima, P., M.L. Lima, U. Kovitvadhi, *et al.* 2012. A review on the "in vitro" culture of freshwater mussels (Unionoida). *Hydrobiologia* 691(1):21–33.

Lopes-Lima, M., R. Sousa, J. Geist, *et al.* 2017. Conservation status of freshwater mussels in Europe: state of the art and future challenges. *Biological Reviews* 92:572–607.

Lydeard, C., M. Mulvey, and G.M. Davis. 1996. Molecular systematics and evolution of reproductive traits in North American freshwater unionacean mussels (Mollusca: Bivalvia) as inferred from 16S rRNA gene sequences. *Philosophical Transactions of the Royal Society of London B* 351:1593–1603.

Lydeard, C., R.H. Crowie, A.E. Bogan, *et al.* 2004. The global decline of nonmarine mollusks. *BioScience* 54:321–330.

Lynn, J. W. 1994. The ultrastructure of the sperm and motile spermatozeugmata released from the freshwater mussel *Anodonta grandis* (Mollusca, Bivalvia, Unionidae). *Canadian Journal of Zoology* 72:1452–1461.

Mackie, G.L. 1984. Bivalves. Pages 351–418 in K.M. Wilbur, A.S. Tompa, N.H. Verdonk, and J.A.M. van den Biggelaar (editors), *The Mollusca, Vol. 7: Reproduction.* Academic Press, Orlando, FL, USA.

Mair, R.A. 2013. *A Suitable Diet and Culture System for Rearing Freshwater Mussels at White Sulphur Springs National Fish Hatchery, West Virginia.* Master's Thesis, Virginia Polytechnic Institute and State University, Blacksburg, VA, USA.

Mansur, M.C.D. and N.M.R. Campos-Velho. 1990. Técnicas para o estudo dos gloquídios de Hyriidae (Mollusca, Bivalvia, Unionoida). *Acta Biolica Leopoldensia* 12(1):5–18.

Matisoff, G., J.G. Fisher, and S. Matis. 1985. Effect of macroinvertebrates on the exchange of solutes between sediments and freshwater. *Hydrobiologia* 122:19–33.

McCall, P.L., M.J.S. Tevesz, X. Wang, and J.R. Jackson. 1995. Particle mixing rates of freshwater bivalves: *Anodonta grandis* (Unionidae) and *Sphaerium striatinium* (Pisidiidae). *Journal of Great Lakes Research* 21:333–339.

McMahon, R.F. 1991. Mollusca: Bivalvia. Pages 315–390 in J.H. Thorp and A.P. Covich (editors), *Ecology and Classification of North American Freshwater Invertebrates*. Academic Press, New York, USA.

McMichael, G.A. and T.N. Pearsons. 2001. Upstream movement of residual hatchery steelhead into areas containing bull trout and cutthroat trout. *North American Journal of Fisheries Management* 21:943–946.

McMichael, G. A., T. N. Pearsons, and S. A. Leider. 1999. Behavioral interactions among hatcheryreared steelhead smolts and wild *Oncorhynchus mykiss* in natural streams. *North American Journal of Fisheries Management* 19:948–956.

Milam, C.D., J.L. Farris, F.J. Dwyer, and D.K. Hardesty. 2005. Acute toxicity of six freshwater mussel species (glochidia) to six chemicals: implications for daphnids and *Utterbackia imbecillis* as surrogates for protection of freshwater mussels (Unionidae). *Archives of Environmental Contamination and Toxicology* 48:166–173.

Molony, B.W., R. Lenanton, G. Jackson, and J. Norriss. 2003. Stock enhancement as a fisheries management tool. *Reviews in Fish Biology and Fisheries* 13:409–432.

Morowski, D.E., L.J. James, and R.D. Hunter. 2009. Freshwater mussels in the Clinton River, southeastern Michigan: an assessment of community status. *Michigan Academician* 39:131–138.

Morrison, P.A., C.M. Gatenby, J. Devers, M.A. Patterson, and R.M. Mair. 2013. *Salvage and Refuge of Target Mussel Species (2004–2012): Mitigation of Bridge Replacement at East Brady, PA*. Final report from the U.S. Fish and Wildlife Service to PENN DOT: Memorandum of Agreement no. 430658 work order no. 1a and 1b.

Mulcrone, R.S. 2004. *Incorporating Habitat Characterstics and Fish Hosts to Predict Freshwater Mussel (Bivalvia: Unionidae) Distributions in the Lake Erie Drainage, Southeastern Michigan*. Dissertation, University of Michigan, Ann Arbor, MI, USA.

Murphy, G. 1942. Relationship of the freshwater mussel to trout in the Truckee River. *California Fish and Game* 28:89–102.

Naimo, T.J. 1995. A review of the effects of heavy metals on freshwater mussels. *Ecotoxicology* 4:341–362.

NatureServe. 2015. *Natural Heritage Central Databases.* NatureServe, Arlington, VA, USA.

Navarro, J.M., and J.E. Winter. 1982. Ingestion rate, assimilation efficiency and energy balance in *Mytilus chilensis* in relation to body size and different algal concentrations. *Marine Biology* 67:255–266.

Negus, C.L. 1966. A quantitative study of growth and production of unionid mussels in the River Thames at Reading. *Journal of Animal Ecology* 35:513–532.

Neves, R. J. 2004. Propagation of endangered freshwater mussels in North America. *Journal of Conchology, Special Publication* 3:69–80.

Neves, R.J., A.E. Bogan, J.D. Williams, S.A. Ahlstedt, and P.W. Harfield. 1997. Status of mollusks in the southeast. Pages 43–85 in Benz, G.W. and D. E. Collings (editors), *Aquatic Fauna in Peril: The Southeastern Perspective. Southeast Aquatic Research Institute Special Publication I.* Lenz Design and Communications, Decatur, GA, USA.

Newton, T.J. and M.R. Bartsch. 2007. Lethal and sublethal effects of ammonia to juvenile *Lampsilis* mussels (Unionidae) in sediment and water-only exposures. *Environmental Toxicology and Chemistry* 26(10):2057–2065.

Nezlin, L.P., R.A. Cunjak, A.A. Zotin, and V.V. Ziuganov. 1994. Glochidium morphology of the freshwater pearl mussel (*Margaritifera margaritifera*) and glochidiosis of Atlantic salmon (*Salmo salar*): A study by scanning electron microscopy. *Canadian Journal of Zoology* 72(1):15–21.

Nichols, S. and D. Garling. 2000. Food-web dynamics and trophic-level interactions in a multispecies community of freshwater unionids. *Canadian Journal of Zoology* 78:871–882.

Nickum, M.J., P.M. Mazik, J.G. Nickum, and D.D. MacKinlay (eds.) 2004. *Propagated Fish in Resource Management. American Fisheries Society Symposium 44,* American Fisheries Society, Bethesda, MD, USA.

Nobles, T. and Y. Zhang. 2011. Biodiversity loss in freshwater mussels: importance, threats, and solutions. Pages 137–162 in O. Grillo (editor), *Biodiversity Loss in a Changing Planet.* InTech, Rijeka, Croatia.

Noga, E.J. 1996. *Fish Disease Diagnosis and Treatment.* Mosby-Year Book, Inc., St. Louis, MO, USA.

O'Brien, C.A. and J.D. Williams. 2002. Reproductive biology of four freshwater mussels (Bivalvia: Unionidae) endemic to eastern Gulf Coastal Plain drainages of Alabama, Florida, and Georgia. *American Malacological Bulletin* 17(1):147–158.

Ortmann, A.E. 1911. A monograph of the najades of Pennsylvania. Parts I and II. *Memoirs of the Carnegie Museum.* 4:279–247.

1912. Notes upon the families and genera of the Najades. *Annals of the Carnegie Museum* 8:22–365.

1918. The nayades (freshwater mussels) of the upper Tennessee river drainage. With notes on synonymy and distribution. *Proceedings of the American Philosophical Society.* 57:521–626.

1920. Correlation of shape and station in fresh-water mussels (Naiades). *Proceedings of the American Philosophical Society* 59:269–312.

Owen, C. T. 2009. *Investigations for the Conservation and Propagation of Freshwater Mussels.* Dissertation, University of Louisville, Louisville, KT, USA.

Owen, C.T., J.E. Alexander, and M. McGregor. 2010. Control of microbial contamination during in vitro culture of larval unionid mussels. *Invertebrate Reproduction and Development* 54:187–193.

Parodiz, J.J. and A.A. Bonetto. 1963. Taxonomy and zoogeographic relationships of the South American naiades (Pelecypoda: Unionacea and Mutelacea). *Malacologia* 1(2):179–213.

Pereira, D., M.C.D. Mansur, and D.M. Pimpoa. 2012. Identificação e diferenciação dos bivalves límnicos invasores dos demais bivalves nativos do Brasil. Pages 75–94 in M.C.D. Mansur, C.P. Santos, D. Pereira, *et al.* (editors), *Moluscos Límnicos Invasores no Brasil: Biologia, Prevenção e Controle.* Redes Alegre, Porto Alegre, Brasil.

Pereira, D., M. C. D. Mansur, L. D. S. Duarte, *et al.* 2014. Bivalve distribution in hydrographic regions in South America: historical overview and conservation. *Hydrobiologia* 735(1):15–44.

Perrier, C., R. Guyomard, J.-L. Bagliniere, N. Nikolic, and G. Evanno. 2013. Changes in the genetic structure of Atlantic salmon populations over four decades reveal substantial impacts of stocking and potential resiliency. *Ecology and Evolution* 3(7):2334–2349.

Piper, R.G., I.B. McElwain, L.E. Orme, *et al.* 1982. *Fish Hatchery Management.* U.S. Department of the Interior, Fish and Wildlife Service, Washington D.C., USA.

Ponis, E., R. Robert, and G. Parisi. 2003. Nutritional value of fresh and concentrated algal diets for larval and juvenile Pacific oysters (*Crassostrea gigas*). *Aquaculture* 221:491–505.

Poole, K.E. and J.A. Downing. 2004. Relationship of declining mussel biodiversity to stream-reach and watershed characteristics in an agricultural landscape. *Journal of the North American Benthological Society* 23:114–125.

Popov, I.Y. and A. N. Ostrovsky. 2014. Survival and extinction of the southern populations of freshwater pearl mussel *Margaritifera margaritifera* in Russia (Leningradskaya and Novgorodskaya oblast). *Hydrobiologia* 735:161–177.

Pritchard, J. A. 2001. *An Historical Analysis of Mussel Propagation and Culture: Research Performed at the Fairport Biological Station.* Natural Resource Ecology and Management Publications. 58. http://lib.dr.iastate.edu/nrem_pubs/58 (accessed September 2017).

Quilhac, A. and J.Y. Sire. 1999. Spreading, proliferation, and differentiation of the epidermis after wounding a cichlid fish, *Hemichromis bimaculatus. The Anatomical Record* 254(3):435–451.

Rach, J.J., T. Brady, T.M. Schreier, and D. Aloisi. 2006. Safety of fish therapeutants to glochidia of the plain pocketbook mussel during encystment on largemouth bass. *North American Journal of Aquaculture* 68:348–354.

Rathke, J. 1797. Om Dam-Muslingen. *Naturhist Selskabets Skr Kjobenhavn* 4:139–179.

Reuling, F. H. 1919. Acquired immunity to an animal parasite. *Journal of Infectious Disease* 24:337–346.

Ricciardi, A. and J.B. Rasmussen. 1999. Extinction rates of North American freshwater fauna. *Conservation Biology* 13(5):1220–1222.

Richard, P.E., T. H. Dietz, and H. Silverman. 1991. Structure of the gill during reproduction in the unionids *Anodonta grandis, Ligumia subrostrata,* and *Carunculina parva texasensis. Canadian Journal of Zoology* 69(7): 1744–1754.

Richman, N. I., M. Böhm, S. B. Adams, *et al.* 2015. Multiple drivers of decline in the global status of freshwater crayfish (Decapoda: Astacidea). *Philisophical Transactions of the Royal Society B.* 370:1–11.

Riusech, F.A. and M.C. Barnhart. 2000. Host suitability and utilization in *Venustaconcha ellipsiformis* and *Venustaconcha pleasii.* Pages 83–91 in R.A. Tankersley, T. Watters, B. Armitage, and D. Warmolts (editors), *Proceedings of the Captive Care, Propagation, and Conservation of Freshwater Mussels Symposium, March 6–8, 1998, Columbus, Ohio.* Ohio University Press, Columbus, OH, USA.

Roberts, A.D. and M.C. Barnhart. 1999. Effects of temperature, pH, and CO_2 on transformation of glochidia of the flat floater mussel, *Anodonta suborbiculata. Journal of the North American Benthological Society* 18(4):477–487.

Rodgers, S.O. 1999. *Population Biology Of The Tan Riffleshell (Epioblasma florentina walkeri) and the Effects of Substratum and Light on Juvenile Mussel Propagation.* Master's Thesis. Virginia Polytechnic Institute and State University, Blacksburg, VA, USA.

Roe, K.J., A.M. Simons, and P.D. Hartfield. 1997. Identification of a fish host of the inflated heelsplitter *Potamilus inflatus* (Bivalvia:Unionidae) with a description of its glochidium. *American Midland Naturalist* 138:48–54.

Rogers-Lowery, C.L. and R.V. Dimock. 2006. Encapsulation of attached ectoparasitic glochidia larvae of freshwater mussels by epithelial tissue on fins of naive and resistant host fish. *Biological Bulletin* 210:51–63.

Schwartz, M.L. and R.V. Dimock, Jr. 2001. Ultrastructural evidence for nutritional exchange between brooding unionid mussels and their glochidia larvae. *Invertebrate Biology* 20(3):227–236.

Seddon, M., C. Appleton, D. Van Damme, and D. Graf. 2011. Freshwater molluscs of Africa: diversity, distribution, and conservation. Pages 92–119 in W. Darwall, K. Smith, D. Allen, *et al.* (editors), *The Diversity of Life in African Freshwaters: Under Water, Under Threat. An Analysis of the Status and Distribution of Freshwater Species Throughout Mainland Africa.* IUCN, Gland, Switzerland.

Shadoan M.K. and R.V. Dimock. 2000. Differential sensitivity of hooked (*Utterbackia imbecillis*) and hookless (*Megalonaias nervosa*) glochidia to chemical and mechanical stimuli (Bivalvia: Unionidae). Pages 93–102 in R.A. Tankersley, D.I. Warmolts, G.T. Watters, *et al.* (editors), *Freshwater Mollusk Symposium Proceedings.* Ohio Biological Survey, Columbus, OH, USA.

Sietman, B.E., J.M. Davis, and M.C. Hove. 2012. Mantle display and glochidia release behaviors of five Quadruline freshwater mussel species (Bivalvia: Unionidae). *American Malacological Bulletin* 30(1):39–46.

Smith, D. G. 2001. Systematics and distribution of the recent Margaritiferidae. Pages 33–49 in G. Bauer and K. Wächtler (editors), *Ecology and Evolution of the Freshwater Mussels Unionoida.* Springer Verlag, Heidelberg, Germany.

Southgate, P.C., P.S. Lee, and J.A. Nell. 1992. Preliminary assessment of a microencapsulated diet for the larval culture of the Sydney rock oyster, *Saccostrea commercialis* (Iredale and Roughley). *Aquaculture* 105:345–352.

Spooner, D.E. and C.C. Vaughn. 2006. Context-dependent effects of freshwater mussels on stream benthic communities. *Freshwater Biology* 51:1016–1024.

Spooner, D. E. and C. C. Vaughn. 2009. Species richness and temperature influence mussel biomass: a partitioning approach applied to natural communities. *Ecology* 90:781–790.

Steingraber, M.T., M.R. Bartsch, J.E. Kalas, and T.J. Newton. 2007. Thermal criteria for early life stage development of the winged mapleleaf mussel (*Quadrula fragosa*). *American Midland Naturalist* 157:297–311.

Sterki, V. 1903. Notes on the Unionidae and their classification. *Nautilus* 12:18–21, 28–32.

Strayer, D. L. 2006. Challenges for freshwater invertebrate conservation. *Journal of the North American Benthological Society.* 25:271–287.

2014. Understanding how nutrient cycles and freshwater mussels (Unionoida) affect one another. *Hydrobiologia* 735:277–292.

Strayer, D.L. and H.M. Malcom. 2012. Causes of recruitment failure in freshwater mussel populations in southeastern New York. *Ecological Applications* 22:1780–1790.

Strayer, D. L. and D. R. Smith. 2003. *A Guide to Sampling Freshwater Mussel Populations*. Monograph 8, American Fisheries Society, Bethesda, MD, USA.

Strayer, D.L., N.F. Caraco, J.J. Cole, S. Findlay, and M.L. Pace. 1999. Transformation of freshwater ecosystems by bivalves: a case study of zebra mussels in the Hudson River. *Bioscience* 49(1):19–27.

Strayer, D.L., J.A. Downing, W.R. Haag, *et al.* 2004. Changing perspectives on pearly mussels, North America's most imperiled animals. *Bioscience* 54:429–439.

Surber, T. 1912. *Identification of the Glochidia of Freshwater Mussels*. Report and Special Papers of the U.S. Bureau of Fisheries.

Tankersley, R.A. 1996. Multipurpose gills: effect of larval brooding on the feeding physiology of freshwater unionid mussels. *Invertebrate Biology* 115:243–255.

Tankersley, R.A. and R.V. Dimock, Jr. 1993. Endoscopic visualization of the functional morphology of the ctenidia of the unionid mussel *Pyganodon cataracta*. *Canadian Journal of Zoology* 71(4):811–819.

Taylor, C. A., G. A. Schuster, J. E. Cooper, *et al.* 2007. A reassessment of the conservation status of crayfishes of the United States and Canada after 10+ years of increased awareness. *Fisheries* 32(8):372–389.

Thomas, G. R., J. Taylor, and C. Garcia de Leaniz. 2010. Captive breeding of the endangered freshwater pearl mussel *Margaritifera margaritifera*. *Endangered Species Research* 12:1–9.

Timmons, M.B. and J.M. Ebeling. 2010. *Recirculating Aquaculture, 2nd edn.* Cayuga Aqua Ventures, Ithaca, NY, USA.

U.S. Fish and Wildlife Service and National Marine Fisheries Service. 2000. Policy regarding controlled propagation of species listed under the Endangered Species Act. *Federal Register* 65:56916–56922.

Uthaiwan, K., N. Noparatnarapom, and J. Machado. 2001. Culture of glochidia of the freshwater pearl mussel *Hyriopsis myersiana* (Lea, 1856) in artificial media. *Aquaculture* 195:61–69.

Uthaiwan, K., P. Pakkong, N. Noparatnarapom, L. Vilarinho, and J. Machado. 2002. Study of a suitable fish plasma for *in vitro* culture of glochidia *Hyriopsis myersiana*. *Aquaculture* 209:197–208.

2003. Studies on the plasma composition of fish host of the freshwater mussel, *Hyriopsis myersiana*, with implications for improvement of the medium for culture of glochidia. *Invertebrate Reproduction and Development* 44(1):53–61.

Utterback, W.I. 1931. Sex behavior among naiades. *Proceedings of the West Virginia Academy of Science* 5:43–45.

van der Schalie, H. 1970. Hermaphrodism among North American freshwater mussels. *Malacologia* 10:93–112.

Vaughn, C. C. 2010. Biodiversity losses and ecosystem function in freshwaters: emerging conclusions and research directions. *Bioscience* 60:25–35.

Vaughn, C.C. and D.E. Spooner. 2006. Unionid mussels influence macroinvertebrate assemblage structure in streams. *Journal of the North American Benthological Society* 25:691–700.

Vaughn, C.C., K.B. Gido, and D.E. Spooner. 2004. Ecosystem processes performed by unionid mussels in stream mesocosms: species roles and effects on abundance. *Hydrobiologia* 527:35–47.

Vörösmarty, C. J., P. B. McIntyre, M. O. Gessner, *et al.* 2010. Global threats to human water security and river biodiversity. *Nature* 467:555–561.

Wächtler, K., M.C. Dreher-Mansur, and T. Richter. 2001. Larval types and early postlarval biology in naiads (Unionoida). Pages 93–125 in G. Bauer and K. Wachtler (editors), *Ecology and Evolution of the Freshwater Mussels Unionoida*. Ecological Studies 145. Springer, Berlin, Germany.

Walker, K. F., H. A. Jones, and M. W. Klunzinger. 2014. Bivalves in a bottleneck: taxonomy and conservation of freshwater mussels (Bivalvia: Unionoida) in Australasia. *Hydrobiologia* 735:61–79.

Waller, D. L. and L. G. Mitchell. 1989. Gill tissue reactions in walleye *Stizostedion vitreum vitreum* and common carp *Cyprinus carpio* to glochidia of the freshwater mussel *Lampsilis radiata siliquoidea*. *Diseases of Aquatic Organisms* 6:81–87.

Waller, D.L., J.J. Rach, W.G. Cope, and G.A. Miller. 1995. Effects of handling and aerial exposure on the survival of unionid mussels. *Journal of Freshwater Ecology* 10:199–208.

Walne, P.R. 1964. The culture of marine bivalve larvae. Pages 197–210 in K. Wilbur and C.M. Yonge (editors), *Physiology of the Mollusca, Vol. 1*. Academic Press, New York, USA.

Wang, N., C. G. Ingersoll, I.E. Greer, *et al.* 2007a. Chronic toxicity of copper and ammonia to juvenile freshwater mussels (Unionidae). *Environmental Toxicology and Chemistry* 26(10):2048–2056.

Wang, N., C.G. Ingersoll, D.K. Hardesty, *et al.* 2007b. Acute toxicity of copper, ammonia, and chlorine to glochidia and juveniles of freshwater mussels (Unionidae). *Environmental Toxicology and Chemistry* 26(10):2036–2047.

Wang, N., C.G. Ingersoll, C.D. Ivey, *et al.* 2010. Sensitivity of early life stages of freshwater mussels (Unionidae) to acute and chronic toxicity of lead, cadmium, and zinc in water. *Environmental Toxicology and Chemistry* 29(9):2053–2063.

Wang, N., R.A. Consbrock, C.G. Ingersoll, and M. C. Barnhart. 2011. Evaluation of influence of sediment on the sensitivity of a Unionid mussel (*Lampsilis siliquoidea*) to ammonia in 28-day water exposures. *Environmental Toxicology and Chemistry* 30:2270–2276.

Wang, N., C. D. Ivey, C. G. Ingersoll, *et al.* 2016. Acute sensitivity of a broad range of freshwater mussels to chemicals with different modes of toxic action. *Environmental Toxicology and Chemistry* 36:786–796.

Watters, G.T. 1999. Morphology of the conglutinate of the kidneyshell freshwater mussel, *Ptychobranchus fasciolaris*. *Invertebrate Biology* 118(3):289–295.

Watters, G.T. and S.H. O'Dee. 1999. Glochidia of the freshwater mussel *Lampsilis* overwintering on fish hosts. *Journal of Molluscan Studies* 65:453–459.

Watters, G.T., M.A. Hoggarth, and D.H. Stansbery. 2009. *The Freshwater Mussels of Ohio*. Ohio State University Press, Columbus, OH, USA.

Welker, M. and N. Walz. 1998. Can mussels control the plankton in rivers?: a planktological approach applying a Lagrangian sampling strategy. *Limnology and Oceanography* 43(5):753–762.

Whelan, N.V., A.J. Geneva, and D.L. Graf. 2011. Molecular phylogenetic analysis of tropical freshwater mussels (Mollusca: Bivalvia: Unionoida) resolves the position of Coelatura and supports a monophyletic Unionidae. *Molecular Phylogenetics and Evolution* 61:504–514.

Whyte, J. N. C., N. Bourne, and C. A. Hodgson. 1989. Influence of algal diets on biochemical composition and energy reserves in *Patinopecten yessoensis* (Jay) larvae. *Aquaculture* 78:333–347.

Wiley, R.W. 2008. The 1962 rotenone treatment of the Green River, Wyoming and Utah, revisited: lessons learned. *Fisheries* 33:611–617.

Williams, J.D., M.L. Warren, Jr., K.S. Cummings, J.L. Harris, and R.J. Neves. 1993. Conservation status of the freshwater mussels of the United States and Canada. *Fisheries* 18:6–22.

Williams, J.D., A.E. Bogan, and J.T. Garner. 2008. *Freshwater Mussels of Alabama and the Mobile Basin in Georgia, Mississippi, and Tennessee*. University of Alabama Press, Tuscaloosa, AL, USA.

Winter, J. E. 1978. A review on the knowledge of suspension-feeding lamellibranchiate bivalves, with special reference to artificial aquaculture systems. *Aquaculture* 13:1–33.

Wood, E.M. 1974. Some mechanisms involved in host recognition and attachment of the glochidium larva of *Anodonta cygnea* (Mollusca: Bivalvia). *Journal of the Zoological Society of London* 173:15–30.

Wurts, W.A. and R.M. Durborow. 1992. *Interactions of pH, Carbon Dioxide, Alkalinity and Hardness in Fish Ponds*. Southern Regional Aquaculture Center

Publication No. 464. Southern Regional Aquaculture Center, Stoneville, MS, USA.

Yeager, M.M., D.S. Cherry, and R.J. Neves. 1994. Feeding and burrowing behavior of juvenile rainbow mussels, *Villosa iris* (Bivalvia: Unionidae). *Journal of the North American Benthological Society* 13:217–222.

Young, M. and J. Williams. 1984a. The reproductive biology of the freshwater pearl mussel *Margaritifera margaritifera* (LINN.) in Scotland I. Field studies. *Archiv für Hydrobiologie* 99:405–422.

   1984b. The reproductive biology of the freshwater pearl mussel *Margaritifera margaritifera* (LINN.) in Scotland II. Laboratory studies. *Archiv für Hydrobiologie* 100:29–43.

Zale, A.V. and R.J. Neves. 1982. Fish hosts of four species of lampsiline mussels (Mollusca: Unionidae) in Big Moccasin Creek, Virginia. *Canadian Journal of Zoology* 60(11):2535–2542.

Zimmerman, G. F. and F. A. de Szalay. 2007. Influence of unionid mussels (Mollusca: Unionidae) on sediment stability: an artificial stream study. *Fundamental and Applied Limnology* 168:299–306.

Zimmerman, L.L. and R.J. Neves. 2002. Effects of temperature on duration of viability for glochidia of freshwater mussels (Bivalvia: Unionidae). *American Malacological Bulletin* 17(1/2):31–35.

Index